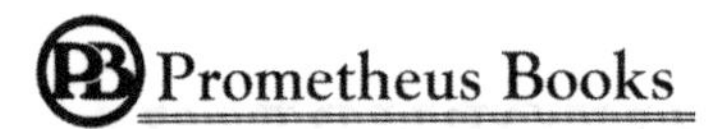

An imprint of Globe Pequot, the trade division of The Rowman & Littlefield Publishing Group, Inc.
4501 Forbes Blvd., Ste. 200
Lanham, MD 20706
www.rowman.com
Distributed by NATIONAL BOOK NETWORK

British Library Cataloguing in Publication Information Available

Library of Congress Cataloging-in-Publication Data
Names: Lesk, Steven, author.
Title: Footprints of schizophrenia : the evolutionary roots of mental illness / Steven Lesk, M.D.
Description: Lanham, MD : Prometheus, [2023] | Includes bibliographical references and index. | Summary: "Steven Lesk, though, after a medical career dedicated to those affected by schizophrenia and a determination to find the answer to its existence, presents a groundbreaking theory that will forever change the lives of the mentally ill. In Footprints of Schizophrenia: The Evolutionary Roots of Madness, Lesk threads evolutionary evidence with neurological evidence, turning the mysteries of our minds into a tapestry of logic. With his breakthrough theory and this unprecedented book, Lesk will invite necessary cultural dialogue about this stigmatized illness, provoke new psychiatric and pharmacological research, and provide unequivocal comfort to those afflicted and affected by schizophrenia"—Provided by publisher.
Identifiers: LCCN 2023014338 (print) | LCCN 2023014339 (ebook) | ISBN 9781633889286 (cloth) | ISBN 9781633889293 (epub)
Subjects: LCSH: Schizophrenia—Genetic aspects. | Schizophrenia—Treatment.
Classification: LCC RC514 .L395 2023 (print) | LCC RC514 (ebook) | DDC 616.89/8—dc23/eng/20230706
LC record available at https://lccn.loc.gov/2023014338
LC ebook record available at https://lccn.loc.gov/2023014339

The paper used in this publication meets the minimum requirements of American National Standard for Information Sciences—Permanence of Paper for Printed Library Materials, ANSI/NISO Z39.48-1992

To my amazing wife, Eileen, who deserves nothing but the world's respect; my brilliant daughter, Phoebe, who is wise beyond her years; and my best teachers . . . my patients

Contents

Introduction

What if we could amble up the cobblestone steps of a high tower and see, perhaps, not just distance but the inexplicable expanse of time? In time's proximity we would see monumental changes occurring over the wisp of the past two hundred years: digitalization, automobiles, the creaking gears of industrialization. Farther off, a group of pelt-wearing, long-haired, bearded men hover around a boulder carving primitive petroglyphs into its broad surface and others decorate caves with crudely painted animals. Farther still, cave people forage for food and scavenge for a vulnerable animal to eat. One glossy-eyed, bearded hulk points toward a fruit bush and articulates a syllable, perhaps "skus," or what would be in English, "berry." His hirsute peers eye him skeptically. He points again, undaunted, walks to the bush, and picks a few, popping them into his mouth. In bemused imitation, the rest eat some, repeating his arbitrary syllable: "skus." Optimistically enthused, he keeps picking and eating while howling the label he's conjured up while the others smile at his senseless blabber, but from that moment forward, they use that sound, "skus," to symbolize that fruit.

That, or something like it, accruing slowly perhaps, was the start of it all, the bomb that detonated evolution's timeworn trajectory forever. A line was drawn. There were animals and along with them hominids (human-like animals)—Neanderthals, Denisovans, and Homo naledi—and then there were Homo sapiens, that unique species of hominin with a special penchant for words. They had a thirst for them, their brains waiting eagerly for a chance to blossom, starving for transformation. Language suited their minds well. Perhaps it was the folding of the cerebral cortex in ways that simplified and accelerated communication.

Perhaps it was some innate oral pleasure in forming symbols, making sounds with their mouths that stood for things . . . Who knows? (The vocal apparatus of Homo sapiens had a greater range than that of Neanderthal, and this may have facilitated an intricacy of speaking that was necessary to promote word use.) Whatever the case, with the onset of language, the massive ship of evolution began a creaky, ponderous yaw, like a barge bound toward a distant buoy. We're still in the midst of that shift, but it's very hard to see something that sharpens over multiple generations. Consider the old Charles Addams cartoon: Two men who look like archeologists are standing in the desert with binoculars and searching for evidence of dinosaurs. One says to the other, "I don't see anything." From our vantage point, however, we can see they're standing smack dab in the middle of a giant T. rex footprint.

Gazing off into the expanse of some six million years from our temporal time tower, we see hominins being chomped on by bears, freezing to death, withering away in a desperate search for food. Closer in, we see people going to college, scientists communicating by computer, mothers in grocery stores conjuring up that evening's dinner. Unlike cavemen, these modern sapiens no longer seemed concerned that their next step could be their last. They lounge in comfortable, heated houses and go to work in cars that practically drive themselves. Whole libraries are stored on a computer chip. Darwin's law of survival of the fittest has been replaced by survival of the most reproductive. Sapiens moved from the jungle to the disco. What happened?

Language—that one sound, "skus," or whatever it was, set the stage for the greatest change in hominin life ever, greater than walking upright, conquering the mystery of fire, or defending ourselves with crudely sharpened sticks. Our brains were drooling for language having no other rickshaw of further growth. Eventually words were syntactically arranged into concepts, and ultimately conscious thought arose when we realized we could talk inwardly to ourselves, which led to conceptualization, to deductive reasoning that set the mind ablaze. The rule of survival of the fittest gave its blessing to the problem-solving ability that came from thinking and sharing those thoughts. How could it not since with its help we neutralized 90 percent of our natural-born predators? Defying entropy, we increased our excitation and complexity. No longer just experiential, physical beings, we became . . . contemplative. Our consciousness, no longer just a hologram of screen images, sounds, internal feeling states, memories, and emotions, became conceptual.

Our minds organized around language like iron filings around a magnet, and with that came a blessed promotion to more-adult ways of thinking. The chemicals that govern our brains, neurotransmitters, and their receptors on the ends of our nerve terminals had to make accommodations. Just as children think differently than adults, primitive man used primitive logic. It wasn't until words and then concepts arose that this maturation occurred. This exponential shift in brain functioning has taken place in the last fifty thousand years . . . a drop in the evolutionary bucket. Fortunately, most of our brains are comfortable with the transformation.

But some are not, and just before their brain fully matures, the old ways of thinking reassert themselves. After all, a six-million-year-old heritage may not disappear so easily. Petulant and left behind, the old way of thinking seeks to regain its former dominance over our central nervous system. In 1 percent of the population it does so in a very avaricious way. We call these people *schizophrenics*.

A much greater percentage of the population (15 to 20 percent) will experience a leaden, bludgeoning major depression during their lifetime (two-thirds of those so affected are women). One to two percent will be bipolar (swinging between depression and a revved-up, carnival state we call *mania*). These too have roots in evolution's course correction, their prototype coming from the hibernation state. Some animals shut down completely in winter depending on their distance from the equator or their evolutionary ancestor's distance. Sensitive to sunlight's yearly modulation, sapiens have also inherited seasonality from plants. There are anxiety disorders like panic and obsessive compulsive, all of which reflect a nod to primitivity, neurotransmitter alterations, and our more simplistic, entropic past.

Getting back to our time tower, we can see the vast expanse of Homo sapiens' trajectory. We're nearly unrecognizable from that skulking cave creature who first uttered the syllable "skus." Was all this planned, perhaps orchestrated by a wizard, or was it some megalithic screwup that was never meant to be? After six million years, we woke up and started speaking, and speaking aided thinking. Certainly the expansion of our brains over time, blessed by evolution, contributed to it. The hollow of our throats could enunciate more sounds than brother Neanderthal's hollow. The changes wrought by all of our knowledge, just in my lifetime, have been exponential. Not all of them are necessarily good, but they are dazzling, hypnotic, seismic.

This book is about how these stunning shifts in evolution's trajectory as we neutralized the forces of adversity that made us easy prey for six million years may have inadvertently created the collateral damage of mental illnesses. The overriding changes in man's brain don't sit comfortably in us all, and holdovers from the past, including the dread of annihilation, still exist in some of us in the form of internal predators. Entropy abhors the organization and increased excitation of our new, shimmering gray matter. Crumbling like a house of cards, entropy helps pull the sufferer's brain back to the old hominid theater that reigned supreme for millennia. One rule seems to be repetitively proven: the primitive seeks to reassert itself. This is primarily motivated by the relentless crush of entropy and our proximity in time to our primitive past. As we will see, the wily brain-salve dopamine is de-suppressed in every mental illness, and in fact, the suppression of dopamine may be Homo sapiens' greatest achievement, setting us apart from our Homo neighbors whom we annihilated. We are still in the midst of this ponderous, creaking turn, and as we sweep through it, these glitches may gradually disappear over perhaps the next, oh, ten to twenty thousand years. At that point there will be no more mental illness, not as we define it. If, from our tower, we could visualize the future, we would know.

Part I

Chapter 1

The Crucible

Each new variety and ultimately each new species is produced and maintained by having some advantage over those with which it comes into competition; and the consequent extinction of the less favored forms almost inevitably follows.

We stopped looking for monsters under our bed when we realized that they were inside us.

Evolution is written on the wings of butterflies.

—Charles Darwin

Whether you believe in creationism or Darwinism, the following formula seems nearly inevitable: protoplasm + mutation x time = human. What does that mean? Living matter (protoplasm) progresses by random error or mutation. That helical molecule called *DNA* that regulates the construction and functioning of cells and ultimately our bodies as phenotypes screws up. Thankfully. For without those blessed mistakes, the changes that modify what proteins cells manufacture would not happen. (Imagine a world where the very first DNA molecule replicated but never mutated. The only life on earth would be that most primitive of cells, the precursor to us all.) And each one of those changes is adjudicated by that greatest of all critics, natural selection. So if a chimpanzee is suddenly born with fuzzy toes that curl a little more completely around branches, he has a distinct advantage over his brethren. His tree-climbing function is assisted, and the next time the chimpanzee is being chased by a leopard he will use those limb-grasping toes to escape by scrambling up the nearest tree a little quicker. When man moved from warm to cold climates, his nostrils kindly enlarged (through mutations) to warm the chilly air he breathed. This clear advantage in frigid climes won the natural-selection seal of approval—at least for those tribes that migrated north. When a gradient is set up favoring one trait over another, natural selection determines the utility of each change (most much more subtle than that) and over mega-bastions of time. The winners rock; those that decrease survival are doomed and quickly perish. And of course, changing environments means changing evolutionary priorities.

Evolution is a sausage grinder and an avid defender of organic matter. Infatuated with the organic, it eschews the entropic inorganic from whence it came, the inorganic lying down passively and obeying the laws of nature without contention. Mother Evolution fights like a scrappy street kid to promote organicity and worships time, honing its precepts to a superlative intensity. The blasphemy of organicity in the majestic universe's vast time without it makes living matter precious to evolution. One might easily question, as I have, why it is so hung up on promoting the organic. Rather than launching into a speculative debate, let's just agree that it is and it fights vigorously for it. Is this another of the basic properties like gravity and entropy? You decide. Perhaps, since it is dearly anti-entropic we should call it *antropy*. Evolution is entropy's greatest enemy. Is time then nothing more than organicity's reward? One might even suggest that time only exists where there is intense conflict between entropy and antropy. (Yet entropy inadvertently aids evolution and hence antropy by favoring the randomness of mutation . . . a process exploited by DNA.) Time only matters to the organic and mostly to us sapiens. A walrus is not concerned with time, but we are, and that is why we label it, measure it, and constantly pay attention to it as our time on earth is so limited. We also have the autonoetic sense of our existence forward and backward in time.

From single-celled organisms to sponges and jellyfish to bilaterally shaped beings with spinal cords to vertebrates, evolution is a grandiose trial-and-error module. Like a car manufacturer, evolution thrusts different models into the market to see what sells. Those that don't are sent to the used-car lots of extinction. (This explains why extinction, or the lack thereof, is the central paradox of schizophrenia. A mutation that causes changes just as the organism is about to leave adolescence and reduces the chances of reproduction [the organism's fecundity ratio] or survival must become extinct, Darwin said. And fast. Well, schizophrenia is not extinct. Any true theory of the illness must explain that, which makes one question the role of genetics in this illness.) Clearly the bilateral-vertebrate model sold most enthusiastically and has remained the sine qua non of organic game plans since then.

That's the crucible. One then could predict that as the crucible simmers its giant cauldron over mega-expanses of time, given the syncopated multiplicity of possible mutations that might arise and their outcome, an animal similar to a Homo sapiens could eventually make its entrance on the organic stage. Millions of mutations have been sent to the used-car lots; millions more have predictably remained to serve living matter well. These are evolution's blessed handmaidens.

Evolution has taken some wrong turns however. Think dinosaurs and their dominant theme: size. In this case, size most definitely did matter. If an alligator was slithering up to bite you and you could just lift up one leg and crush it, you were protected. It stands to reason then that when it came to predation, bigger was better. An elephant could defend itself against a fox, and a dinosaur could defend itself against an elephant. Evolution tumbled down the rabbit hole of size in a big, brassy way. The law of the crucible was to affirm those mutations that promoted survival, and size most decidedly did.

But evolution went a little too far. The amount of food that dinosaurs required was massive, perhaps almost unsustainable, and yet the dinosaurs, many of them herbivores, survived until, according to one theory, about sixty-six million years ago when a fiery asteroid smashed into earth in the Yucatan Peninsula, blackening the skies and burning up a lot of that food in hellish conflagrations. T. rex and his friends dropped like huge, charred flies. Either they suffocated, starved, or got scorched directly by the explosion. Here we see a course correction made nearly inevitable by circumstances that on the whole seemed perfectly reasonable. Evolution created a monster that folded under its own weight with a little nudge from its friend the asteroid.

So what did evolution do? Brutal necessity breeds change, and size, no longer the winning mantra, was replaced by something else. But what? What could possibly beat size in survival power? In its search for a new, steamy paradigm, evolution moved upward. While body plans were important, the behavioral repertoire was more decisive. Cunning became the new deity as brain power proved its worth.

And this wasn't evolution's only course correction. Something else intervened more powerfully than an asteroid. Language.

* * *

Evolution exerts its effects not just on the body but on the mind. The bundle of nerves that control the organism also fall under natural selection's adjudicative spell. Our central nervous system's very beginnings inform the motion of that process. We can assume that initially primitive organisms, including single-celled ones, just had simple reflex arcs. These promoted survival by dragging the mini-organism swiftly from the arc of danger and urgently toward nutrients and hospitable environs. Gradually those primitive organisms like jellyfish, sponges, chordates, and vertebrates blossomed more complex central-nervous-system repertoires, the opposite of entropy's desire. Options multiplied beyond forward or reverse. At some point a vague sense of intention arose and the organism made choices . . . where to go and what to do. Striated (or voluntary) muscles were developed as the capacity of the organism for movement increased and mute, autopilot decision making began to take root.

Freud used the word "ego" to denote the command center of the mind, particularly humans' minds. "Normally control over the approaches to motility devolves upon [ego]" (Freud 1957, 215). The primitive ego sucked up the functions of purveyor of this intentionality. Captain Ego juggled intention and choice . . . informed choice. Needing detailed information about the world at large, perceptual systems sharpened, presenting ever-refined, incoming stimuli (sight, sound, touch, smell, taste) to the organism, and as they sharpened they improved survival directly, forwarding better info to the individual. Information was the golden commodity of evolution. The goal was an early warning system, blaring the alarm of predation and a recognition of incoming environmental torpedoes and opportunities, and the sooner the better. Like Captain Kirk, the ego became the commander of this skittish ship, orchestrating crystal perception and intention's lively dance: "the self preservative instinct, which must be assigned to the ego" (Freud 1957, 224). It was the ego's job to protect and promote the organism. (Remember that Freud was a neurologist. His mental theories arose out of a deep understanding of the central nervous system.)

In the service of survival, memory started to pitch in its two cents. A mental deal-maker, it was added to improve the odds of any decision, comparing incoming data to past experiences and solutions and near-tragic missteps. Now decisions could be influenced by what had happened before. Learning from your mistakes was no mean feat and a monumental boon to success. If a gecko remembered that when it looked under a mossy rock it saw a snake, it could then avoid that terror the next time. Or maybe if it remembered where it had stumbled on a luscious, edible worm, it might saunter back for a potential feast. Elated, natural selection certainly gave the thumbs up to memory storage and retrieval. What an upgrade! All of these advances worked in mellifluous concert to improve the odds of survival as the mind differentiated into groovy individual parts with unique assignments. The animal kingdom was surging gloriously in mental abilities under the guidance of advanced orchestration by the central nervous system and natural selection's harmonious nod of approval over eons of time.

Eventually a vague quality called *conscious awareness* developed. Avoiding the mystical, exactly what conscious awareness is remains elusive, and dusty, ponderous tomes have been written about it. Primitive man and animals could then perceive something and have a sense of recognition and *self*-awareness. I am, therefore I am. Godzilla or a rock, predator or brunch, the organism learned to divine the difference . . . wordlessly. At what point conscious awareness barged onto the scene is unknown, but it was certainly before hominins (man and pre-man). Freud understood that just visualizing something was a very imperfect form of awareness. For millions of years humans had what one might describe as an experiential consciousness that was good enough to at least limp us onward. We saw stuff, smelled burning meat, and touched animal pelts and each other, and those perceptions delivered to us the tools we needed to scrape our way forward through the jungle terror. But at some point we became cognizant that we exist over blocks of time forward and back. For example, I'm aware that I've been typing and writing for the past hour and will be for the next. The term "autonoetic" is applied to this.

Certain basic functions like breathing, peristalsis, and heart rate were relegated to automaticity using smooth muscle (as opposed to striated) long before we Homo sapiens arrived. This gradual march forward is quite typical of any developmental schema. (We wouldn't want to have to consciously tell our hearts when to beat, our lungs when to expand.)

This primitive brain organization exists today. It existed in pre-verbal man for six million years before language. It is the stuff of animals, cavemen, young children, and users of mushrooms and other psychedelics. And we revisit it nightly in REM sleep. We were physical, experiential, Neanderthalish grunts, glorified animals, through millions of years of evolutionary time using what Mother Nature gave us to scrape our way to survival.

So then what?

Like an evolutionary balloon, the brain swelled like a wet sponge. For millions of years it was a wordless container, an experiential vessel filled with perceptions, feeling states, memories, and intentions. It was like an octopus-eyed computer sitting inside a closet and humming away. The mind worked silently on autopilot, pulling as many clever strings as it could to promote the survival of the individual. Sitting on a mossy rock, Mr. Neanderthal pondered his next move wordlessly, giving his brain the hefty energy and blood supply it demanded. Fluid streams of impulses, feeling states, memories, intentions, and fears coursed through his brain as it adopted a plan of action. In that unpredictable horror show of an environment, that daily schema, of course, had to be the safest and most likely to succeed if it was to promote survival. Otherwise like his brethren he'd be ground back through the sausage maker of evolution to hitch a ride to the dust bin of earthly inorganic matter, entropy wining again.

And we humans, the most physical of beings, ambled around on our own steam. No fridge, no 7-Eleven, we got up in the morning and would have to grub around in search of delicacies for ourselves and our family. If you were in the mood for wild boar, you ran, homemade spear in hand, to where you thought you might possibly find one, and anything you managed somehow to kill in the hunt, you carried. If a man needed a mate, he found her and chased her down. No matter what his impulses, he simply gratified them crudely, as best he could. Children of nature, we were physical beings with an experiential mind but one that had unconscious, coursing fluid in it, thought streams, that worked silently to promote our precarious welfare. (One might assert that most of the thinking we do now is still this same, pre-verbal cogitation that humans engaged in for six million years while the layer of conscious contemplation wanes thin.) Morality was not our focus so much as survival. Morality would come later with language that groomed cooperation and ultimately civilization with its lock on predation.

Ego, the orchestrating Oz or Captain Kirk of the mind, took control of intentionality. Catering to an endless demand for gratification, the ego did whatever it could to carry out the demands of the individual in a most harrowing, unforgiving habitat. This was the crucible. With the help of the rest of the mind, including conscious awareness and memory, it guided the gruesome caveperson to potential theaters of gratification. So for example, a hunger state presented various enticing options. How had it been solved in the past? Slipping into gear, the brain assisted us in finding the most efficient and hopefully safest solution for this endless gnaw. Even so, sometimes the best-laid plans vanished under a tiger's tooth or a cabal of hungry hyenas.

But evolution needed to appoint a deputy as it is only equipped to indemnify genetic events (mutations); it can't rule on micro-behaviors. The ego had been granted the mystical wand of intentionality. These complex protoplasmic entities behaved with fluctuating purpose and made their own decisions. Educated by brutal experience, some behaviors were random and exploratory while others were pointedly goal directed, and some were more adaptive than others. Lady Luck played a big role in the outcome.

So evolution, anticipating Pavlov, approved a system in which sanctioned behaviors that followed the laws of natural selection were rewarded with positive reinforcement. Good doggy, here's a little treat. Rotten, dangerous behaviors were denied a reward or were possibly even punished by a painful stimulus. Evolution needed a sugary chemical that would function in the brain as its surrogate, as a seductive reinforcer. One could consider the brain then as a second evolutionary messenger once removed. Painful stimuli had already been established to inform the organism that some damage had occurred because it was important for the protoplasm to know it had been injured. It would also serve organic matter well to know it had carried out the mandates of Madam Queen Evolution successfully. The reward chemical chosen by evolution was . . . dopamine!

Dopamine is a neurotransmitter, perhaps the most famous of them all. A cornerstone of mental illness etiology, this talented transmitter rules the gelatinous brain-matter kingdom. Technically, a neurotransmitter is a chemical released by one nerve cell to cross the gap between that cell and another nerve cell (a gap called a *synapse*) and trigger a response in that neighboring nerve cell. The neurotransmitter is welcomed into its specific receptor in the receiving cell with great fanfare just as a key enters a lock. It then gets reabsorbed into the first cell from whence it came, which requires another lock-and-key receptor, this time called a *transporter*, to drag it back in. Released to encourage appetitive behavior in a Pavlovian-reward paradigm, dopamine had a special place in evolution's heart. Reproductive behaviors were of course at the top of the list. As I said, evolution cackles with glee in the face of behavior that promotes the continuity of organic matter. It is entropy's sworn enemy. Other survival-adaptive endeavors like food gathering received dopamine's (and evolution's) consummate and immediate nod as did aggressive and protective stuff. Dopamine screams self promotion. This deputized reward system served Homo sapiens (and other animals) well. Evolution's handmaiden, it did its bidding and got evolution's highest seal of approval, le Prix d'Evolution. As in Pavlov's experiments, dopamine was the conditioned stimulus that made certain behavior more likely (LeDoux 2020).

The simple art of cooking with fire shortened the bowel's digestion time and ultimately its length leaving more blood supply for the brain, which used about a quarter of the body's energy, hogging the incoming energic reserves. In short, protoplasm's nervous systems went from simple reflex arcs to lumbering, decision-making eggs, albeit silently[Q: ?]. Evolution, reeling from the T. rex fiasco, had done itself proud this time. Although size still mattered and a large gorilla could rip a man to shreds (and often did), cunning's mystical prowess proved insurmountable, and a sapiens with a bow and arrow might just fell the hairy ape from a distance. Walking on two limbs not four allowed us to stand taller with an advantageous view, carrying tools and weapons in our hands across grassy plains. (Many of these changes in habits took millions of years. They were a result of climatic shifts, dietary changes, and sometimes just plain luck coupled with mutations.)

So man was still not at the top of the food chain (Harari 2015), and the life of early man was brutal. Existence was not only not guaranteed; disaster was nearly an expectation. If a Neanderthal (one type of hominin, the general category of human-like animals, and another species of the genus Homo) for example, broke his hip, he was basically finished. Dragging himself to his cave if he could, he waited to die: no food, no water, no crutches, no Percocet. Scratching his way to survival, everything demanded an aggressive, no-holds-barred intensity. He was used to terror and pain; he knew victory and was all too aware of gruesome defeat. And in this regard dopamine served him well. Humans have more dopamine receptors than any other animal by brain weight (Vernier et al. 1993). It provides energy, aggressiveness, and maybe even optimism. It suited the physical, experiential nature of our ancestors.

Our primitive hominin had no use for morality, civilization's mortar. If an urge or impulse made itself known to him, he gratified it. One and done. No doubt the need to procreate was high on nature's list of sanctioned endeavors. In what manner primitive man did this we don't exactly know, but we can assume there was capitulation of the unwilling, more often force. (On the other hand, there were little family units that no doubt served the purpose of protecting the pregnant woman and then the small hominin until it could fend for itself.) The legacy of that gonzo, amoral, prehistoric lifestyle lives on in us all. At some point in our infancy we draw a curtain across this terrorizing mental state and put it behind us. Freud called this *infantile amnesia*, and it was said to happen around age five. We forget the gruesome terror and brutality of that drama. Civilization is based on pushing it aside, perhaps delusionally, to create a quiescent civility. When we regress to psychosis, that door is opened up again and often reveals feelings that have been submerged since infancy: the door to the psychotic id. Is civilization based on a delusion of safety? Perhaps, or possibly just the need to maintain a sense of security that promulgates itself, in fragile pose, like a ballerina en pointe too long. The maintenance of this civilized state of calm has much to do with the suppression of dopamine.

Evolution shifted. Size still mattered but in another way: brain size. (Things that once were external tend to move internally in the developmental schema of things.) The forces of adversity in the jungle environment overwhelmed: predators large and small, illnesses, injury, starvation, and extremes of weather all played a role in the unforgiving crucible of natural selection. Eons of time allowed these intense challenges to simmer. Eventually man's brain reached a critical mass and organization. "The relative size of the entire cerebral cortex (including white matter) goes from 40% in mice to about 80% in humans" (Hofman 2014, 4). The brain of Homo erectus was about 900 cubic centimeters compared with the 1350 cubic centimeter Homo sapiens brain (Leakey 1994). "The brain accounts for 2–3% of the total body weight but it consumes 25 percent of the body's energy" (Harari 2015, 9). The brain, cunning, became the new evolutionary god. In concert with the body it proved its silent worth time and again. On a righteous path, evolution patted itself on the back, having finally redressed the dino fiasco.

But hold on . . . Evolution had no idea what it was in for.

Chapter 2

Along Came a Spider

The Sixth Sense

We must reject the idea that the faculty of articulate language resides in a fixed, circumscribed point.

—Pierre Paul Broca, 1861

It was not planned. Certainly evolution did not foresee this turn of events. The brain of a Homo sapiens reached a critical mass and configuration. Like the critical mass of plutonium that leads to nuclear fission, the onset of language led to an explosion far greater than the thunder of an atomic bomb: the big bang, if you believe in language's abrupt arrival as some do. And of course the atomic bomb is a perfect example of what we were capable of once our language birthed contemplation. With the help of good old Captain Kirk, the ego, the brain organized around this new principle: words.

Initially words were labels, and they may have been object labels or syncretic, combination labels of actions and things. For example, a word may have developed for the action of throwing a rock at a moose, like "moosculating." Starting a fire with a stick may have been mashed into one word or the joy of eating a trout. However, words gradually assumed specificity, modifying labels into more refined categories . . . it's not just a rock, it's a big rock; it's not just a flower, it's a bright red flower. That's not just a hippopotamus over there; it's an angry one about to charge . . .

Run like hell! All of these divisions and subdivisions and descriptives and qualifiers encouraged subtle mental differentiations that were previously impossible without, of course, speech.

Suddenly a dividing line was drawn between what was delivered to consciousness and what was held back by our friend the ego. Before words, the consciousness of man was experiential. The dull Neanderthal took note of sensory inputs from his or her eyes, ears, skin, nose, and internal feeling states. (The pathway to consciousness remains the sensory channels. When we think, we talk to ourselves in the form of internal speech.) Our ego was primitive and had difficulty differentiating self from others, inner from outer. Just as children may believe that clairvoyant Momma knows what is on their mind, primitive man was just as underdeveloped. Freud, in denoting his classic view of the mind, did not take pre-verbal man into account as definitively as perhaps he should have. Freud's metapsychology included the famous unconscious, preconscious, and conscious areas. Sans language, those cubby holes are nonexistent. He did ask, however, how something could become preconscious, and the answer was "By coming into connection with the verbal images that correspond to it" (Freud 1957, 213). In short, a flapping fish barrel of words began to coalesce in the sapiens' mind, a small vocabulary. Obviously not all could be in conscious awareness at the same damn time. But words allowed the ego to construct the mental apparatus that led to adult thinking. The ego became the word processor supreme, the street cop, waving some forward, stifling others. The ego gained the important task of fish finding but not only this. Along with it came reality testing, syntax, and hierarchical organization of incoming stimuli. Captain Kirk earned a huge promotion. The demarcation between one's self and others was hardened, baked to a golden crisp. The traffic cop orchestrated the dance, the flow, the pattern of shoulds and shouldn'ts in the mental theater.

Initially a thin layer of preconscious labels accrued. Language's hypnotic influence masticated and intensified the mental actors. As vocabulary expanded, so did the preconscious (that part of the unconscious just next to what is conscious or aware.) Like guests on the *Tonight Show* waiting patiently in the green room, ethereal thought-blobs approached their word labels in the preconscious just before they took the platform of awareness. Modifiers, descriptives, and feeling states were all wordified. The mind moved from diffuse to categorical specificity, a process unavailable to the pre-verbal, primitive (mental) organization. Words were organized into sentences that obeyed the rules of syntax (grammatical mandates). And the importance of syntax in organizing our minds should not be underestimated. Some believe that it is an innate trait of a sapiens. A master fisherman, the ego's task was to reach into the fish barrel of verbiage and pull out just the perfect terms to decorate those blobby, fluid thought precursors. Fido thinks this way, pre-verbally, as did man, pondering silently, wordlessly, but actively nonetheless. Their smaller animal brains work to promote their interests as did the brain of Joe Neanderthal. And of course our brains still think this way for the most part I would assert. Only a slim fraction of our thoughts are promoted to the level of conscious awareness. Syntax, with its subject, object, verb, adjective, and so forth, was one system that advanced brain development. Before reaching consciousness, our ego commander dusted sentences off, reality testing them and hanging them out to dry on the eave of awareness. Eventually this whole process reached a state of automaticity.

Thus a fancy tripartite organization imposed itself on the mind. Words were scooped from the preconscious, organized into sentences, inspected with a magnifying glass by the ego's reality testing function, then silently delivered through the auditory channels. (We consider the forebrain, most likely the prefrontal cortex, our gelatinous mind's seat of consciousness.) The act of delivering concepts has many implications, like a crazy Amazon driver who does a dance as he drops your toaster oven on the front porch. For one, a chemical called *BDNF* (brain derived neurotrophic factor) is secreted, and this stuff's MiracleGro for the mind. It promotes dendritic arborization, the little nerve-cell branches fanning out like tree roots. Without that the forebrain has a potential for atrophy and especially if it doesn't get enough incoming entertainment.

And BDNF is euphorigenic. A gaping mouth hungry for info, the forebrain needs this input much as a muscle needs exercise to prevent it from withering. When deprived, in a desperate effort to avoid deterioration, it starts creating its own cinematic input. This is the stuff of dreams at night, while during daylight hours they are hallucinations. Mental illness therefore interrupts the smooth flow of nutritious activity to the forebrain, which needs that input. Restoring this incoming frequency is always the treatment goal and we do so by toning down or blocking dopamine, directly or indirectly. As we grow up, we reexperience this evolutionary brain intensification as the rules of talking go from diffuse, infantile idiosyncrasy to specific, directed, informational, and communal (Werner 1948, 289). What the adult says, the rest of us understand. With children and schizophrenics on the other hand or psychedelic users, sometimes their responses are so unique to their own experiences that others just don't get it because that person is still in the process of solidifying the congealing brain organization that language bestows. In a great leap (over fifty thousand years or so) the ego took a 180-degree turn from the caveman's primitive organization, hurtling our cognitive development forward in ways that were unachievable before we spoke. And it was the ego, the commander of this gelatinous gray matter, that got us out of Dodge.

So our ego is at this point stuffing the forebrain with stimuli of a conceptual nature. It is now behaving like any other sensory organ (eyes, ears, etc.). Only instead of a Bach concerto or a twilight panorama, it delivers the stimulus of organized concepts (not sounds, not sights) to our conscious awareness, and it has become, so to speak, a sixth sensory organ. Language, at some critical juncture, moved from labeling and symbol creation to thought when we realized we could talk inwardly to ourselves. And with this monumental progression, language created several nifty advantages for the mind of humankind. Now Freud's metapsychology with its conscious, preconscious, and unconscious areas could be realized and defined in a word-juggling context that before language was a no-go. With the ego we have a sense of identity, something that shreds to confetti if we drink ayahuasca or take LSD. Defining boundaries, our beefy egos drew a sharper demarcation between our self and the rest of the world. Our cognitive abilities multiplied like horny rabbits. The ego had the function of a monolithic word processor. Scooping the best words out of the preconscious part of the mind, it hammered them together syntactically following the rules of grammar (which we all learn from Mrs. Petry in sixth grade) into gaily expressive sentences. Like Sherlock Holmes, it inspects those sentences using the rules of adult logic, presenting them to consciousness like a neatly bowed gift delivered by FedEx. It is then consciousness's role to open that gift and contemplate it. Happy birthday.

Words hit the spotlight and danced. Not only did language serve the purpose of expression, but it procured the mind's expansion. And conveniently, when we wanted not to remember something, all we needed to do was dissect it from its sticky word label and leave it unattended, dangling alone in the preconscious. Or even further, as the unconscious was an even deeper burial, we could hide it down the stairs and in the basement. Ideas, memories, or impulses could be banished there like a bastard child, rarely if ever to be seen, miles from the words that describe them. Civilization demanded that certain impulses never see the light of day. Impulses like sleeping with Mom, killing your neighbor, and smearing feces all over were sent downstairs. Like unwanted relatives, they rarely got an invitation to the reunion. Sometimes they got feisty and threatened to bust the door down. This, Freud noted, created mental symptoms as the mind struggled to keep gross Uncle Fred from the appetizers and the other guests.

And words were the mind's ticket to a new dance with dopamine: suppression. The mind learned this clever feat of legerdemain from the motor system. We need to suppress dopamine, the ubiquitous neurotransmitter, to attain ballet-smooth, coordinated movement. A magician's sleight of hand or the complex hurl of a knuckle baller both rely on this suppression. This brings dopamine into a delicate balance with another neurotransmitter, acetylcholine, in the motor nerve tract called *the nigrostriatal*. But the mind lacked the tools to do this in the other dopamine tracts, mesolimbic and mesocortical, as if we lacked the limbs to carry out this conceit. Language gave them to us. We could now suppress dopamine in those tracts, clearing the mind space, gating out unwanted intrusions from our mental playground, supporting the mighty ego and assisting the complexification of our minds to adult rules. We were now freed to enter the process of control that led to clear thinking, and we did not need dopamine's reward function as dearly since we had our own mystical evaluator in the form of thought. Our minds were desperate recruits waiting for the tools to enact this precious symphony. Thus when dopamine de-suppresses, either in a regressive way or under the influence of chemical stimulation or in sleep, the mind space becomes flooded and disorganized. Hallucinogen users, meth addicts, and schizophrenics report this regularly . . . as do manics. How do we treat this? With chemical agents that brilliantly block dopamine's receiving dock, their receptors. But I'm getting ahead of myself.

A crude analogy might be as follows: a man stands on a rock at a stream casting a net into the gurgling waters. He is looking only for an engagingly unique kind of trout. He readies his net and when he spots the perfect fish scoops it out of the water, examines it, and, if the fish is suitable, stores the flapping monster in his creel. He does this until his creel is brimming with the slimy swimmers. He then organizes his creel neatly until he has the right combination and hauls it back home where his wife and kids are waiting arms crossed for their next feast. He hands it to his wife who looks skeptically over the menagerie and makes an assessment of their suitability for cooking. Depending on her judgement, they might have a delicious dinner or a smelly garbage pail.

In this analogy, the man is the ego. He is there scooping out the flapping fish words from the preconscious stream. He, the ego, is choosing just the right ones. Then the ego organizes the words and delivers them, now formed into syntactical sentences and concepts, to his wife and kids (consciousness), who pass judgment on them.

And remember, without our new trait of dopamine suppression, the fish could only leap randomly up at the man perhaps accidentally falling into his creel. This state of affairs might happen when the ego is really weak and unable to do much organizing—a state of de-suppressed dopamine. Suppressing dopamine strengthens the ego and keeps out unwanted mental intrusions. Or perhaps he had a few beers on the way to the stream, or maybe took some acrid ayahuasca or licked a croaking Sonoran desert toad, resulting in de-suppression of dopamine, and puff, dissolution of the ego. It's a flooding tsunami of ungated input the mind can't get a handle on. In fact, one could define the major mental illnesses as word diseases. What I would call *linguistic deterioration* is a hallmark of schizophrenia and mania (although there are differences: syntax is often maintained in mania). Why? Because our fisherman ego may become impaired and not be in control of the word choice or the reality testing process as our chemical friend dopamine is not suppressed. These symphonic components must work together like clockwork.

It was words and the utilization of them by the ego that got us out of our prehistoric Dodge in the first place. Without words we'd be glued to the primitive mental organization cave humans like Mr. Neanderthal had for millions of years before language. This is another indication that mental illness is directly woven into language's jurisdiction. Again, in order for the mind to organize words and offer them a clear stage for inspection, *we must suppress dopamine*. You see certain trends emerging in all of this, trends that are indicative of the problems we note in mental illnesses, particularly dopamine nonsuppression, treatable with, as I mentioned, dopamine-blocking agents, putting the brakes on this ubiquitous chemical. But other medications hamstring dopamine too, albeit indirectly.

Not only that, but words created a second reality so to speak, a scaffolding around the structure of our feelings and perceptions. No longer was a state of hunger, let's say, the only thing connected to memories of starvation. Past solutions to hunger, locations of food, preferences—now the *word* "hunger" magnetized all of those dangling connections. A unique secondary plane of reality built up around the framework of words and their remembered stirrings. Something called *laterality* happened as the brain expanded on one side (the dominant, left side for most of us right-handers) to accommodate language and all of its glittering accoutrements.

But wait, *there's more!* This ego, newly minted and deputized, became our ticket out of the primitive organization. Sapiens went from the jungle to the disco, and primitive brain organization, that infantile mental state 1.0 by which pre-verbal caveman operated, was ultimately replaced by the post-verbal, complex adult 2.0 operating system of modern humans. The Captain Kirk of our brain ship *Enterprise* delivered us from the impulse driven, amoral, experiential-physical state of primitive souls to the civilized, law-abiding, moral contemplative beings that we are today.

Religion was born as was government, institutions of higher learning, jurisprudence. All of this soared around our mind's new operating system 2.0. In the spirit of social cooperation we learned to live side by side. And another of Freud's concepts (we never get *too* far from Freud when we think of the human mind or Darwin when we contemplate evolution), the superego, rewarded us for law-abiding behavior. Embracing the golden rule, now we had a good feeling inside when we did the right thing, and this is the basis upon which civilization rests. Wrongdoing feels bad.

The power upon which man once operated had been usurped by this new sixth sense, word processing, telescopic wonder. Where did Captain Kirk get all his strength? From that primitive operating system itself that used to rule our minds. Kirk muscled up and took from the primitivity of our minds his supremacy in language's intricate ballet. Robbing Peter to pay Paul, anything that strengthened the ego took more from the old 1.0 and gave it to the new-and-improved 2.0. An impairment in the ego's power now reflects mental difficulty, a symptom of coming infirmity. When the ego is crippled, watch out. Schizophrenics may be described as suffering from ego weakness (Tausk and Feigenbaum 1992), but this would have to be true for bipolars (manics) and unipolars (major depressives) as well, and they experience lack of gating. Out of chaos comes order. And order is supplied by the ego with the help of dopamine suppression in the context of our spiffy new operating system. When order is lost, the chaos of entropy—which lurks behind this process, seeking disorganization and lower energy states—again reigns. All of this goes along with dopamine de-suppression as the fulcrum of organized thinking painfully reverses. A schizophrenic breakdown or an LSD trip use the same pathway. We see things like linguistic deterioration as syntax vanishes, flooding of the mind space as gating dissolves, and phenomena like blockage of thought or perhaps even word salad. Their opposites were originally unified in a global dopamine-suppressive growth pattern reproduced in all of us as we mature.

And our egregious friend the ego grows from certain activities that accrue in the strabismal playing field of youth. We learn to suppress dopamine and learn the art of contemplation. Those brain fertilizers

BDNF and VEGF accrue, enhancing the mind's own enrichment. Reinforced by parenting, schooling, discipline, rules, and socialization with its steady partner cooperation, ego grows. Dragging us out of our entropic primitivity, all of this propels us from primitive thinking to adult, modern conceptualizing. The child no longer says, "The moon is a big fat man that will eat you!" The rules of logic crystallize into adult conventions while the mind becomes the most anti-entropic, or antropic, substance in the universe by gaining organization and energy.

Developmental psychology tells us a lot about this. As a basic rule, any process of development goes from diffuseness, disorganization, and lack of structure to specificity, hierarchic organization, and a structural basis (Werner 1948). When the high school band leader blows the whistle, the kids line up and start marching in unison. Children's minds, hallucinogen users' minds, our sleeping minds, primitive people's minds, and schizophrenics' minds all lack specificity. The vagueness that is indicative of early development reigns over childhood's incentives. There is also an idiosyncratic quality to a child's speech. Children respond in ways that are unique to their own experiences as do schizophrenics in their brazenly autistic way. I once asked a schizophrenic man, "Why does the sun come up in the morning?" His reply was "Tomorrow." I looked at him blankly. When I requested explanation, he said, "Didn't you ever see *Annie*? The sun'll come up, tomorrow," and he burst into song. Instead of addressing the question from a factual point of view, he simply referred to an idiosyncratic association he had to my question. Adult thinking is universal and understandable by all who speak-a da language. It is a core symptom of schizophrenia, this regressive type of autistic thinking typical of children—or at least we evolutionists think so and most psychiatrists do as well.

One more thing! Freud taught us that there are two types of thinking, primary process and secondary process. Akin to daydreaming, *primary process* is unstructured, a rollicking roller coaster ride. We stare at the lamp, our minds scatter in a million diamond shards of connection and glances of memory. Okay, but this is not the type of thinking that suppresses dopamine, hardens the ego, strengthens our adult conceptual rules, or secretes BDNF or VEGF. Sorry. Our Neanderthal brethren indulged in it only without words. Before language, our participation in our own mind's clockwork ordinances was nil, nothing, nada. The ego was not exerting any structure or control over it. Psychedelics, for example, blast the ego from its moorings and put us in a state of evanescent daydreaming. Poof! The ego vanishes. They do this by stimulation of a type of serotonin receptor called *5HT2a* (the "5HT" part just stands for serotonin, five hydroxytryptamine). Suck the ayahuasca cup and the ego shreds; all we can do is sit back and let our mind lead the way. There's no gating, no dopamine suppression—frankly, there's just the opposite.

Now the next type of thinking is called *secondary process*. This is the type of deliberate, effortful mulling that warps into problem solving and focuses on a chosen topic. You get the idea. Deliberateness. That requires the participation of the ego and . . . us. We all have these two mental modus operandi, Jekyl and Hyde: the six-million-year-old trail wandering, which for most of that time went on without language but now employs words and visualization and rambles on autopilot. And there is now a new, ego-driven jet stream that is less than fifty thousand years old and that goes on with muscular ego guidance as an active, participatory effort since it takes decided effort to think. (That effort it turns out is in overcoming entropy.) Two modes, one ancient and one new. The ego evaporates, leaving us in a state of passive, experiential consciousness when, as I keep saying, we take psychedelics, a hot topic in the psychiatric kitchen. We're dazzled by the mind's shift to a more primitive thought paradigm. When we actively think, using the muscular ego as a tool, we strengthen it. We reach solutions to problems, we explain and understand, and we actively suppress dopamine, gate out unwanted intrusion, complexify the very rules of thinking itself, and puff up the brain with the indelible growth euphoriants BDNF and VEGF, a whole smorgasbord of positives that were unavailable to the Denisovans or Neanderthals. These ethereal brain chemicals cause the nerves to expand and branch out. Like a muscle, if you use the brain, it enlarges; if you don't, it degenerates. Use it or lose it. Is it any wonder that schizophrenics are in danger of atrophy of the cortex? No. Their mind structure is reverting back to pre-verbal times after they've mastered operating system 2.0. There is some dopamine suppression involved with coordinated movement. This is referred to as "body ego." The forebrain needs conceptual input just as a muscle needs to be exercised so the analogy between brain tissue and muscle is quite apt. Both are prone to dissolution without use and input. One could rightly say that thinking is muscular activity. However, this loss of brain function is not inevitable since schizophrenics are still able to use secondary process thinking albeit with difficulty, offering a possible avenue for restitution of their mentation.

Secondary process thinking requires dopamine suppression. In order to clear the brain space or theater of cognition within the mind, dopamine must be suppressed. When schizophrenics break down, dopamine painfully de-suppresses and their minds are flooded with a tsunami of unprioritized stimuli. Their ability to think is vastly inhibited although not impossible. A dose of dopamine-blocking agent such as a Thorazine tab of maybe fifty milligrams can clear their mind as it restores the gating function. Now the seat of contemplation is free to rumble whereas before it was overwhelmed. The preferred input of conceptualization had been chaotically usurped by in-rushing stimuli that left the forebrain in a starvation mode akin to sleeping. The brain interpreted this as a state of sleep and began to self-stimulate with dream-like hallucinations. And the analogy between hallucinations and dreams is quite apt. Both are creations of the forebrain routed through the sensory pathways under conditions of sensory deprivation. When the primitive organization takes over in schizophrenia, the forebrain interprets this as a state of sleep and begins its daytime hallucinatory dreaming as a means of self-stimulation.

* * *

This is just the very start of where we're going to go. To grasp the influence of evolution and its changes on psychiatric infirmity we have to understand what happened when the symphony of words arrived on the scene and not just the radical modifications we inflicted on Mother Nature that were wrought by language but in our brains. Clearly the psyche of a modern-day accountant is not the same as the psyche of Mr. Neanderthal, that spear-wielding, pelt-wearing ogre (although he may have had many gentle qualities). And what we see in mental illnesses reverses some of those modifications.

So let's review some of the mental changes that took place when Mr. Neanderthal went from the brutal daily drama of jungle survival to that accountant with his earbuds in his ear and his computer in his briefcase as he hops into his electric car to take him to the office but stopping no doubt to grab a latte on the way.

Symbol formation became our forte, and we sapiens do it well. Our minds sizzled at the options, taking to it like an addict to dope, our psyches buzzed with the enriching potential. Like Helen Keller in the movie *The Miracle Worker*, we dropped everything and made the startling connection, and it was electrifying. It pioneered our mind's transformation. And it went like this.

Our brain lateralized. The left side swelled to accommodate the enormity that symbol formation, meaning language with all its implications, was.

Next, words became the tornadic center for feeling states. For example, hunger's vast implications now had a label around which solutions, memories, dangers, and implications all coalesced. We erected a pile of ashes into a new configuration, an alternate universe. Taming and in a sense rejecting our natural environment we created our own physical geodome in which to exist. Homo sapiens have put Mother Nature at arm's length with the help of words. Of course, for some six million years or so we were in a perilous engagement with the predators she exposed us to so it would be natural to mistrust her.

At this point, evolution's deputy, dopamine, could now be canned since we could engage in crystal-clear decision making of our own, and clarity of contemplation, a clear brain space, required dopamine's suppression. We needed to put that bad boy in his settled place once and for all. Conceptualizing took over as cavepeople crossed the line into meditative contemplation. A new model of brainism surfaced on top of the old. Dopamine suppression gated out the unwanted stuff, fed the ego and its identity and self-awareness module, fertilized the mind, and complexified, differentiated, and structured the clockwork organization of our hungry psyches.

Yes, all of that in our new operating system 2.0. What we see when dopamine un-suppresses is a return to OS 1.0. It's a loss of the ability to make a point, or apointilism, with linguistic deterioration and a tsunami of unwanted debris flooding the brain space, including hallucinations due to loss of the gating function and sensory deprivation. Perhaps our greatest accomplishment, the suppression of dopamine, set us apart from all other species of Homo. Evolution, knowing a good thing when it sees it, welcomed it with a smarmy, sloppy embrace. As for survival of the fittest, we gradually tamed our multitude of predators, and our lives morphed into survival of the most reproductive as an expectation of longevity settled in.

Yet it should not surprise you that the diseases we might relate to evolution are mostly dopamine driven: Parkinson's disease, schizophrenia, attention deficit disorder (ADD), Alzheimer's disease, stuttering, Tourette's syndrome, and more. We can add to these those ubiquitous mood disorders major depression and bipolar mania and toss in as well common anxiety disturbances, obsessive compulsiveness, panic, generalized anxiety disorder, and so forth. Laying the seed of them all when it flows backward from suppression, dopamine is the bad boy of psychic distress whose de-suppression is behind every mental illness. It's the pin on which everything hinges. The suppressive process reverses course and opens the door to an exquisitely painful reassertion of the dominance of dopamine that occurred for millions of years before language because nearly every medication we use today to reverse mental illness suppresses dopamine directly or indirectly.

Ego, meanwhile, in celebration of its scintillating orchestration efforts, got the promotion of a lifetime and with it a strengthened sense of identity. This word-processing dynamo yanked the words out of the fish barrel and delivered them to Madame Forebrain. Voila, a whole new paradigm of mental modus operandi settled in: contemplation! In a reward switch, not dopamine but BDNF and VEGF graced our shores as the Pavlovian conditioning stimulus. Thinking reinforced mind expansion, enriching our minds with dopamine suppression and complexification of mental apparatuses. We now had the ability to construct hypotheses from preliminary data points. This paradigm of modus operandus 2.0 is rattled by every mental disturbance known to man, and every de-suppression of dopamine mucks it up. Imagine our accountant at a zoo with his earbuds and iPad sitting on a bench and staring through a fence not at a gorilla but at Mr. Neanderthal. The brute would sit on a rock and stare back at him in wordless observation. The accountant would think to himself, "Who is this guy? He is me, but what does he experience in his daily life? Does he think? No, he can't."

So beyond a lack of words there is a lack of brain organization within. The ego organized our brains around language and vice versa, and the conductor of the symphony grand in doing so got a huge promotion. When the ego weakens (in concert with dopamine de-suppression), the whole framework of our mental apparatus melts, distorts, limps, and sputters, and the primitive organization gains strength again.

Now the accountant thinks to himself, "I wonder what it would look like if I took an operating system from the 1980s and put it in my MacBook Pro?" Some things work on a primitive level and others not at all. That's a good analogy for what happens when a schizophrenic's brain crumbles.

Evolution, as we've seen, made a course correction after the dino fiasco. Natural selection morphed to sexual selection. Survival of the fittest yielded to survival of the most reproductive as civilization with its new rules embodied a more humane courtship process. Embracing the memes, evolution took note of society's customs and mores, and those have a lot to do with choosing a mate. The horrors of nature's adversity that enslaved us for millions of years went bye-bye.

And remember all of this took place in the last fifty thousand years or so. If we represent the approximately six million years that hominins have been around by laying three yardsticks end to end to represent those years, the last fifty thousand years would be about nine-tenths of one inch! And these changes didn't take place as soon as language appeared on the scene. It took thousands of years for all of this to gestate and more to proceed from speech to internal thought. The crucible worked its slow magic, interacting with our changing brains to promote conceptualization and cognitive enrichment. The primitive brain organization that ruled man's mind did so for 107 of those 108 yardstick inches. The momentum of that time is certainly with the primitive organization and not our new and improved 2.0 Cadillac, and at the same time, entropy seductively favors simplicity and the disorganization of the old ways.

Length of time living one way isn't the only reason to pay attention to who we are. Evolution takes place in a context. We should pay heed to the conditions in which protoplasm and ultimately humans evolved because those are the conditions that humans adapted to via natural selection. Man evolved in warm climates. He ate whatever foods he could find nearby, and he hunted. He was nothing if not a physical being. We evolved within Mother Nature's parameters, and those mutations that helped us survive within them got natural selection's seal of approval. And we're still getting used to this radical acclimatization. While I'm not suggesting we live like cavemen or follow a paleo diet, we need to pay homage to our lengthy primitive past. Our bodies, for instance have a need to move and were designed to do so. You don't take a tank, for example, that is designed for war, and use it to run errands downtown or to approach the takeout window at McDonald's. A sedentary lifestyle is anathema to what humans evolved to be.

We need to sweat more. Humans moved around in warm climates generally, which led us to lose our fur, and although there was some migration northward, activity in warm air makes one sweat. (Incidentally, the farther one goes from the equator, the greater the suicide rate. We evolved near the equator. There may be a connection there.) Sweating may have been a much greater excretion paradigm than we're now used to. Go to a sauna or a steam room. There are studies that show that using them reduces sudden death and cardiac mortality. Some toxins are stored in the fatty tissue of the skin only to be eliminated as we sweat. Along with sweating we need to replace fluids with good old water. Drink from the stream. Dietarily, ancient humans no doubt ate occasional meats but usually foraged around for edible plants and fruits. Not on a time schedule like a modern office dude, cave guy ate anytime he found something edible.

Sitting eight hours a day may take more of a toll than we realize. We were not designed for that. On the other hand, evolution helps us adapt to these modern demands. Those whose bodies are more tolerant of sitting have no doubt already been favored by natural selection.

If you build something to meet certain environmental demands, it flourishes in those situations, but if in the last inch of our three yardsticks we ignore those demands, our bodies and our minds will not be as prepared for them. These uniquely recent demands become, in a way, spanking new predators and novelties that will gradually change evolution's priorities. Evolution crawls forward by the establishment of gradients, and those who conform most closely to the gradients' requirements survive. The others don't. It makes sense to pay heed to our past even though the demands we place on our bodies (and minds) are now vastly different. We are still in the midst of the course correction that's happening all around us, but we don't see it. We do notice it when things screw up the process. Illnesses can be considered evolutionary misfires if they do not meet the criteria of Darwin's rules regarding genetic principles that usually lead to extinction.

Which brings us to the point that this book will focus on. I've enunciated the vast changes that language brought to our minds. We have new demands in complexity, functioning, and expectation. The modern world is more of a Freudian one than a Darwinian one, I might say. By this I mean that language helped erect Freud's complex metapsychology. Modern man functions within that metapsychology: the ego, id, and superego and the preconscious, conscious, and unconscious forces that rule us. Sexual repression has no doubt lessened since Freud. The mental changes listed above have all taken place in the past fifty thousand years or so. The barge of evolution is making its creaky course correction, but it is still doing so and not all of us are yet able to accomplish it. Perhaps one day, maybe ten or twenty thousand years from now, 100 percent of us will fall into the mode of modern thinking and none of us will be repossessed by our primitive past. But for now we are not quite there, and our primitive past, favored by entropy, will haunt some of us more than others. Don't underestimate entropy's yearning for the simplicity, low energy, and obedience of our inorganic roots as a tornadic force in the return to our old operating system 1.0, a force that, like gravity, never rests.

The primitive organization I've been referring to, in some ways analogous to Freud's id or what some might call a *paleocortex*, is in all of us. For the vast majority of us, the ego drags us forward, leaving behind much of the primitive organization's influence and robbing its power. Most of us move from childlike thinking to adult thinking, from idiosyncratic speech to universally understood communication, from the physical and experiential to the conceptual, and all of this is due to the mystical blessings of language. The cage bars between the accountant and the grunt Neanderthal are the ego. The ego becomes a syntactical word processor and ultimately a sensory organ designed to deliver to us a conceptual symphony through the timeworn auditory pathway for our consciousness's delight. And contemplation strengthens the ego as well as the forebrain as it suppresses dopamine, allowing us to construct new hypotheses from the ashes of data points. All of this is in the context of speech. The ego strengthens and sets up a barrier to the forces of the primitive organization. With dopamine suppression we can gate out those annoying, intrusive, unwanted stimuli. If the ego weakens (as in hallucinogen ingestion, schizophrenic breakdown, or sleep), this can open the door to a resurgence of the primitive organization along with a torrential dopamine flood. Let's begin to explore what all of this means.

So far we've touched on what happened to us over the past fifty thousand years or so since language and not just in neutralizing the gruesome forces of adversity that surrounded us but in our minds. What grew out of language was an entirely eccentric beast, one perhaps unintended by greedy evolution. Had evolution gone down another crooked highway? Like T. rex, will language and its ramifications turn out to be another ponderous misfire? Was any of this meant to be? As we stand in the giant footprint from the Charles Addams cartoon looking for signs of our brutal, jagged past, we can only conclude that we are not who we were. It's best to define as accurately as possible what those changes were and to make note of evolution's screwups, the few of us who have been either left behind or called back to a prehistoric state of mind that existed long ago for millions of years in a land we no longer inhabit. We're not in Kansas anymore, Toto. Evolution's baby may be cranky and disturbed, but overall it's far better off than it was before. Change creates glorious advances and at times collateral damage. Suicide, entropy's greatest victory as it drags us back to the inorganic, and evolution's most dramatic failure, may be one of those problems. Illnesses like schizophrenia, Parkinson's disease, juvenile diabetes, ADD, Alzheimer's disease, Tourette's syndrome, stuttering, and others may be some of evolution's side effects along with major depression and bipolar illness that are also offshoots of our Neanderthal past, the common thread of entropy and dopamine de-suppression. We no longer live in the treacherous jungle of our ancestors, and these newly minted brain songs and bodily readjustments will fit some and destroy others. Perhaps in the future we will settle in comfortably as we distance ourselves from entropy's righteous call.

And then there are the problems that accrued from man's flash of new talent. Pollution, dietary disenfranchisement, sedentary lifestyle, the absence of green nature in our city landscapes, the sense of alienation that takes place among some . . . in these incidental ways we've outsmarted ourselves. We have landed men on the moon while millions here on earth are starving. This is not to say that these problems were all created by progress but that the process remains in flux and that we are still in the midst of it.

Chapter 3

Neanderthal's Dream, or What Would Freud Say about Dopamine?

We have already cited the most important of these postulates . . . The nervous system is an apparatus having the function of abolishing stimuli which reach it, or of reducing excitation to the lowest possible level.

—Sigmund Freud, *Instincts and Their Vicissitudes*

Our primitive Neanderthal patriarch careened about with a brain that we all still nurture deep within. Referring to it as the id, approximately, good old Freud dissected its role in dreaming. Others might call it a paleocortex or the remnants of an ancient mind that did not so much vanish as get overgrown, like a compact boulder laden with deep green moss. This primitive organization as I call it intrudes its ugly head in the thought forms of children, the hatter-like staging of dreams, and the discourse of schizophrenics. Steeped in survival, morality had no seat on the bus, and our cave guy wanted nothing more than to see another day, fill his stomach, and procreate his genetic uniqueness into some craggy future. This primitive director sits inside us all, now buried by the force of ego's promotion and our adult ways of considering things brought to us by linguistic intransigency. Speaking invented us in our transformational costume, which each of us reinvents as we march through adolescence. With the aid of education, study, thought, and discipline, the conductor waved his baton to a new tune, leaving the Neanderthal theater behind with its entropic trough of diffuseness. We trudged from physical and experiential to the contemplative beings we are, and that ignominious transformation delivered to us the world. Yet we find time to revisit old caveperson's mentality at night when alone, paralyzed, and cut off from the day's blaring inputs, our mind harkens back to this six-million-year-old dreamscape bathing in entropy. Nowhere perhaps more than in dreams did Freud's metapsychology come to life.

Words performed their symphonic magic on our souls, weaving subtle complexities into our brains. Our mind began marching to the buzz of words, and under the spell of words, our psyches solidified, differentiating into working categories, juggling a plethora of word options into syntactical productions. Words served as a potent genie of a litany of partitions and subdivisions. Words birthed mind, and mind birthed civilizations since socialization demands a complex strategy, and civilizations require rules, golden rules that ultimately foster taboos. For example, incestual relations were forbidden, which, incidentally, served the dictums of evolution. The goal was to prevent inbreeding, which could lead to weakening of the gene pool. Civilization, humming on the engine of cooperation, forbade certain indelicacies. So there was a time before formal civilizations when clans existed, casual groups of loosely connected caveman families that hung together. In such families it was not so uncommon for the young males to rise up against the elder male leader, sparks flying, and kill Pop, castrating him, in order to claim the women for themselves. This brand of chaos needed burial if civilizations were to survive, and Mother Evolution gleaned the survival value of working together quickly in her keen wit. In the spirit of going along to get along, this mode of rebellion needed a funeral swift and deep. And with our newly minted mental categories, the unconscious afforded us a quaint graveyard. Taboos could be separated from their word labels and then dropped into the depths of that graveyard.

How did this happen since our Neanderthal friend had no need to forbid such impulses? This has much to do with Freud's theory of neurosis, which posits that without internal prohibitions there can be no neurosis. (I might add that without language there could be no neurosis as well. This gets to the question of repressing unaccepted impulses for to do so, words must be separated from them—part of the mental complexification that we discussed in chapter 2.)

Freud, a neurologist, set about understanding the oddly behaving patients that were referred to him. In those days hysterias blossomed like sunflowers in a summer field: patients who appeared to have neurologic illnesses like paralysis of the legs, blindness, imaginary pregnancy, and other phantasmic conditions. After careful examination Freud shook his head and concluded that there was no neurological basis for any bit of it. He yearned to help these mysterious sufferers but had no idea how to go about it since fixing people in a totally new category of illness provided, to say the least, mega-challenges and immense opportunities. In order to prevail, Freud had to create from scratch a whole new architecture of science called *psychoanalysis*, which he created brick by brick from the ground up. He was a quintessential bulwark of a thinker. When all was said and done, he had created a spiffy new theory of the mind with its metapsychology, a method of treatment, a new terminology (including a little thing called the *unconscious*), and a whole lot more. Let's see if we can piece together some of his most germane views about dreams accurately and perhaps where he could have gone a little farther back in time. (By the way, if you haven't read Freud's *Interpretation of Dreams*, give yourself a challenge and go read it. Even if you disagree with everything Freud says, you will have benefited from stretching your mind with this book. And if you only have time for one chapter, choose chapter 7.)

We were talking about inner taboos threatening the pandemic of cooperation that glued society's bricks together. In treating his patients, Freud found that they were blocked while entertaining certain impulses, hidden in the unconscious, that he discerned were expressed metaphorically. Using his newly minted technique of free association, he found patients inevitably worked their way back toward these logjams of forbidden urges. The weird, neurologic-like symptoms he encountered gratified those impulses in a symbolic compromise while at the same time keeping them closeted in the shadowy unconscious. As he became more confident in his views, he questioned why these impulses had to be so hidden. As he tacked together the framework of his metapsychology, he realized the mind was divided into conscious, preconscious (which, as you remember, is where words attach to their objects), and the graveyard of the unconscious. The impulses these hysterical patients were struggling with were deeply submerged like dead submarines in their unconscious while striving for expression. Making patients aware of them was exactly what he needed to do. Yet he met invariably with a mighty, burgeoning resistance as his hysteric patients were far from ready to confront these corpses upon which their symptoms were based. He found similar themes in these patients: sexual impulses toward a parent or sibling, an urge to smear feces that he related to the cleanliness urges of obsessive compulsives, a desire to commit patricide or fratricide, and so forth. A whole host of unseemly ghosts. Eventually he had to ask himself, unwanted as these impulses might be, why did they need such extreme burial in the unconscious?

Natural selection turned its eye from these horrifying deeds since they didn't specifically reduce one's struggle for survival. The outposts of patricide, fratricide, and incest didn't offer specific survival drawbacks to the cave human. Yet under civilization's accrual, they became uniquely abhorrent internalizations, invocations of another era, to the point where now they are as deeply buried in our minds as in a crypt. As early civilizations formed, there grew the need to suppress certain actions that seemed to recur over and over. (Lest we forget, civilization was a natural selection winner. Large groups banding together were much more prepared to defend against even the largest predators, promoting survival. Longevity was much more likely, and food, water, and other basic necessities could be provided within the confines of this larger, group-given cooperation. Unlike photosynthetic plants, hunter-gatherer animals need to expend energy to gain it. For the first time civilization corrected that flaw.)

Freud hypothesized that in early civilizations, the adult sons of the clan's father or leader occasionally killed and castrated him in order to have the women for themselves. (Certain primates have a rigid structure in which the dominant male gets all the women for himself. The other males are left to have intercourse here and there when the dominant male is distracted. Yet it isn't reported that they rise up and kill the dominant male.) When this happened, presumably the word got around to other clans, inducing a sense of horror. (Of course this required language.) This horror turned into strict prohibitions against these rebellions (in the service of civilization), and the prohibitions led to forbidding this kind of behavior. This potent forbidding ultimately became internalized in the subbasement of our minds. These buried taboos were what Freud hypothesized were causing the mischief in hysteric and other patients' equanimity as they pushed for gratification. His job, uncovering and guiding the patient to see these locked-up taboos, was one of dueling with thousands of years of repression. (When we use this word we refer to the mind's insistence on keeping something out of the awareness of the individual. This was the exact force that Freud knew he had to help his patients overcome. When they were successful, with the help of what he called transference, their symptoms generally abated.)

The Oedipus myth is a classic example of internalized prohibition. Euripides wrote his famous play about this. After finding out he has slept with his mother and killed his father, Oedipus pokes his eyes out. Eyes can be interpreted as a metaphor for testicles, and Freud knew that his male patients ultimately had a dread of castration with a rough equivalent of this for women. But the impulse and the punishment were buried way outside the patient's awareness. When impulses such as this became exceedingly strong, a symptom might arise that entwined the individual's need to gratify it with their need to remain unaware of it. As a simple example, a boy with a strong urge to kill his brother might end up in a fight with a schoolmate who happens to look like his brother or has a similar name or other superficial characteristic. (This brings up the whole question of childlike logic. It is based on superficial similarities and idiosyncratic meanings. This kind of primitive logic dominates symptoms in mental illness.) The fight gratifies the impulse while maintaining its disguise. Or a man who lives with his ailing mother might become unconsciously motivated toward an Oedipal relationship with her. He might then develop a tingling in his hands and a paralysis in the legs. The latter prevents him from entering her room while the former is a substitute gratification feeling. The mind finds a way to both gratify (with a sacrificial lamb so to speak) and prevent the impulse from true enactment while keeping it far from our conscious awareness. It does so with the help of childlike logic.

We also invoke certain symptoms or even full-blown diagnoses as bleating, sacrificial lambs to forestall a worse condition. For example the self-cutting or head banging of some borderline patients might be viewed as an attempt to derail a blunt schizophrenic coup by offering it a crumb. Major mental entanglements might be viewed as an attempt to sidestep a schizophrenic collapse by finagling the process down another less-onerous path.

In sleep we float through layers back to the lowest possible state of energy. The body knows how to heal the cuts and savagery inflicted upon it during daylight's encounters. The brain does so by returning to the lowest state of excitation, entropy's desire. This is the state of mind we had before language, that primitive organizational simplicity before we climbed the ladder of linguistic complexity, defying entropy, each night returning to that detumescence. We lie reposed in a state of sensory deprivation, muscles paralyzed, with no light, no sound, and little touch, taste, or smell. In this vacuum, fearing atrophy, the frontal lobes self-stimulate, replaying the day's insults only redoing them with a polished outcome. Through sleep layers, dopamine flows backward in a river of de-suppression while the ego shreds and the gating of stimuli, our psyche's muscular roadblock, is lost. Our dreams are designed to self-stimulate within the context of entropic revery, embracing the primitive organization. We go back to the brain of our grumpy Mr. Neanderthal, and his dreams were exactly like ours only with less disguise.

So for example if one of us might dream of killing his father, after moving through layers of repression, by the time the dream reaches the consciousness of sleep, it hits the stage as follows: you are playing chess with an old man who suddenly grips his chest and collapses. Or to be more obscure, you dream of a clock (representing Father Time) that suddenly starts turning backward and then explodes. You get the idea. The dream uses that childlike logic we just talked about and is camouflaged into nearly unrecognizable form with symbolic accoutrements. You can make up your own tantalizing dream distortions. In dreams the forebrain sends hallucinated scenarios through the sensory pathways to consciousness to relieve some of the atrophic sensory deprivation it experiences. These miniseries don't become conscious until they surge through the sensory pathways, mostly visual, some aural. One could surmise that consciousness arose through the sensory inputs that the brain then interprets in a field of conscious awareness since Mr. Cave Guy was nothing if not experiential. At night with minimal sensory input, the brain, hungry for it, creates its own, dreading atrophy and whittling its way to the valley of deep sleep, its entropic wishes' delight, while gratifying the need to right the daytime's wrongs. When the primitive organization takes over in schizophrenia, the forebrain interprets this as sleep and begins waking hallucinations. This happens in the context of a regressed mental organization and super-charged dopaminergic state woven together in a ballet of forces, restorative and recreative, to rewrite the day's events in bearable form.

Now you can see the parallel between creative neurotic symptoms and dreams. They are both camouflaged like deer hunters and gratify an impulse while maintaining their anonymity in the deep crypt of the unconscious. In their imaginative redress, they return the brain to a low point of excitation, Freud's pleasure principle. Some dreams are disguised more than others, and there are clearly recognizable ones. These are dreams that are farther from societal taboos or insistent door-breaking replications of past trauma. And what about our Mr. Neanderthal? What did he dream? Well, going back to a time when there were no societal taboos, no societies at all, there was no imperative to disguise the dreams that he manufactured and no language. His dreams were a direct depiction of his primitive impulses and an attempt to gratify them and preserve sleep while also rewriting the script of his jungle failures. Freud's forbidden scenarios had much to do with pre-civilization's early genesis. Neither the Neanderthals nor the Homo sapiens had yet to experience these taboos fashioned for proper society. When we as civilized adults dream, we work our way back to the primitive organization that humans knew for millions of years before language and civilization. Then we use our unconscious and power of imagination to camouflage the dream in imaginative ways.

I'm ready to assert here that the deep unconscious is just a place where words are dragged away kicking and screaming from their word objects or targets. Why not, since it's effortless enough to just undo what's going on in that other part of the brain, the preconscious? Words tacked onto their glittery targets became the sine qua non of consciousness. Before that? Visual, auditory only . . . the sensory input channels made for a consciousness of experience. Eyes wide shut, Mr. Neanderthal, the experiential guy, heard and saw, but while seeing may be believing it's not a certainty of knowing. For all intents and purposes Mr. N knew not one thing more than Rover knows when his food dish is being filled. Yes, he gallops over, a fur ball of excitement, salivating, and then eats. Feeling full he goes back to his dog bed and chews an old bone, content that he has eaten after all. It was words and ultimately thoughts that consciousnessed us. We started using labels, then indexes, then icons, then symbols—and language is nothing if not symbol formation. But symbols, born from icons born from indexes, started to coalesce around concepts, and this is where the fireworks began. The transition from labeling all sorts of things, to thinking . . . the real birth of consciousness when we realized we could talk inwardly to ourselves, and with consciousness, the unconscious was born. We magnetize our brilliant contemplations with words and send them through the auditory channel presenting them to the dorsolateral prefrontal cortex, at which point we really do know them. Whatever preconceptual stream of thought blobs exist, they're nothing without words. Tack on a few words chosen by Mr. Ego syntactically, and suddenly we have a brilliantly inspired concept. In short, the whole Freudian metapsychological framework hinged on words. Freud forgot to go back one stop farther to the pre-verbal.

One could say Freud specialized in reconstructions of our moment in time when civilization was getting its legs. It was a thorny transition between cave dweller and our next-door neighbor grilling steaks in a silly smock that reads "I do nothing without wine." Those plays became civilization's crossroads. Euripedes' *Oedipus the King* took center stage with a starring role, the ego, once again. It was our ego that multiplied in strength with the mastery of Oedipus. A little boy dissects away his wish to replace Dad and have Mom all to himself, and as ego strengthens, the end result is an admiration for Dad and a loving but distant concern for dear old Mom. Instead he seeks a partner out there in the world of high school drama, a substitute for Mother, and buries those impulses deep away in the bowels of the unconscious. In Freud's day this format for unconscious impulses occasionally came back to haunt some in the form of hysterical symptoms. These were the very symptoms Freud learned to unearth with his inventive new theory and the therapy of psychoanalysis.

Sleep is the salve to our daily challenges, the buffer to our waking missteps. It lures us to REM and then stages its productions where the primitive organization of our prelinguistic brethren claims the stage. But for us the play is disguised with the wrong costumes, the sets altered and disconnected, the message skewed. In schizophrenia, the same thing swarms the mind, de-suppressing dopamine and dispensing with its gating function, and the brain, interpreting it all as a waking sleep, engages its hallucinatory productions. Echoing voices reverberate with messages of utmost disparagement as the primitive organization takes wicked aim at its enemy the ego. The schizophrenic enters a world of ghostly visions and mumbled voices, a waking sleep as petrifying as any horror film. Abstract paintings, our dreams are as jumbled as a broken funhouse and in need of camouflage, something Mr. Cave Guy had no use for as civilization had yet to issue its demands for cooperative behavior.

Assuming we came from a primitive organization that served humans well for some six million years or so, *that hallowed entity is in us all!* No surprise, it sits like a purring cat (for *most* of us), a remnant of our past. Other brain structures have grown up around it like towering cypress trees in a forest that rises above a stubborn, indifferent boulder that has rested there for six million years or more. Something called lateralization happened. When speech appeared, the left side of our brains expanded to accommodate all that language is. The prefrontal lobes process the ideas presented to them by the conceptual ego, and dopamine suppression facilitates a clarion brain space. The brain is highly complex and a major puzzle still today, but for most of us it hums like a Maserati most of the time. But for some it doesn't, and that has a lot to do with what it was for some six million years before language.

Any transition that involves intrapsychic changes, heavy mental contortion, may also induce a passionate derangement in others. The road from pre-civilization to civilization gave birth to neurosis in certain repressed cultures. With its ankle-length, billowing dresses and covered skin, Freud's stodgy Vienna fit that bill. And what of schizophrenia? The transition from pre-verbal to verbal, the complex changes neurochemical and behavioral wrought by language are still sweeping across the land in an evolutionary frenzy. It's happening all around us; we just can't see it as it inches forward over generations like a giant invisible platypus. Evolution needs to drag us farther from primitivity and entropy's Pied Piper seduction to be free of it at last. Then mental illness will be but a gleam in history's eye, a footnote of collateral damage.

* * *

Not only do we have to have a foothold in Freudian psychology, developmental psychology, and the laws of physics, anthropology, and evolution to grasp these concepts, we need to know an awful lot about brain chemistry. We've talked a bit about dopamine, the neurotransmitter that served humans well for some six million years (and other species long before that) and still does. Evolution gave it a starring role as a reward chemical. But we've evolved to be less experiential and physical animals and more conceptual, going from the jungle to the disco, from uncivilized to civilized. We've seen how this affected and strengthened our egos and how we have put jujutsu layers of defense and repression against our primitive organization's impulses. During sleep it seems as if we drift backward in time through these layers like Alice falling through the looking glass and revisit the primitive organization once again. In REM stage, impulses can be dressed up, camouflaged, and sent to our sleeping consciousness while we're in a state of paralysis, sleep kindly paralyzing us to a state of submission. This state defocuses the body and entirely focuses on the mind while taking power from our conductor-friend the ego. (Remember that motor activity suppresses dopamine and in sleep we want to avoid this.) We also reach a lower state of energy and revel in a restorative, entropic trough, the brain marinating in a bath of lowest excitation. There are numerous implications in all this mostly relating to language. We use concepts delivered to our consciousness by a strong ego that pulls energy away from the primitive organization defying entropy in the *cohones*. So what does all this have to do with dopamine?

Quite a lot. As I've said, we have more dopaminergic neurons per brain weight than any other animal (Vernier et al. 1993). Dopamine served our Neanderthal ogre well. It provided a reward (and still does) for activities that promoted survival like eating and sex. It provided aggression and brain stem activation and even euphoria. (As I sit here eating a piece of chocolate and drinking some cocoa, I can feel the rush of dopamine surging in my brain. Caffeine and theobromine, the more common chocolate stimulant, indirectly pump the release of dopamine from certain nerve cells by blocking a receptor in the brain called the adenosine2/D2 heteromer. D2 is a dopamine receptor, while adenosine causes drowsiness.) Dopamine no doubt enabled our caveman to survive. Other chemicals in the brain serve other purposes.

Not only caffeine but all stimulants promote dopamine, and dopamine is a brain neurotransmitter that bounces back and forth from the end of one nerve terminal to the front of another like a chemical ping-pong ball. Between the nerve cells is the synapse, a liquid-filled chasm. On the far side of that channel are receptors (in the receiving nerve end) that dopamine fits neatly into like a key in a lock. When unlocked by that chasm-crossing dopamine key, the receptor causes the receiving nerve cell to fire. The dopamine that is still swimming in the synaptic divide gets recaptured by a transporter on the first nerve's surface like a weary fisherman dragging his net back to shore. This guy drags the ping-pong ball back and strong-arms it kicking and screaming into the cell that released it in the first place. From there it may get chomped up by other chemicals and enzymes, one of which is called monoamine oxidase. (I know, these *names*. But I'll simplify as much as I can. If you're a nerd like me, you love this stuff.) Or it may get stored in a vesicle (a bag within the cell body that contains it . . . This also requires a transporter). This baggy vesicle can burst open on command from the cell nucleus and hurl its gooey contents back into the synapse.

Every chemical classified as a stimulant works by boosting dopamine directly or indirectly—every single one including caffeine, which, by the way, is no small player in the stimulant grab bag. Caffeine is added to street drug stimulants to boost their effect, and high doses can kill you, so beware the energy drink. The average cup of coffee has about 120 milligrams of caffeine. At about 400 milligrams the thunderous, adverse effects begin to override the stimulant effects. Withdrawal from caffeine can be quite uncomfortable with headache and flu-like effects hanging on for weeks of pain. (By the way, caffeine constricts your cerebral blood vessels, and when it leaves your body, your blood vessels then dilate in a bounce-back fashion causing headaches or triggering migraines.) When we consume our daily caffeine fix, it's to prevent the creeping withdrawal syndrome much like an alcoholic who drinks in the morning to nullify the jitters. Caffeine combined with booze is a particular no-no as the sedative effects of the alcohol are not well perceived. Caffeine is an arousal drug and a motor stimulant and it's swimmingly addictive, all you caffeine addicts.

Other stimulants like methamphetamine (a most dangerous drug), ecstasy, cocaine, and so-called bath salts are also dopaminergic in effect. The stimulant-type medications used for ADD like Adderall and Ritalin are similarly dopaminergic, and of course controlled substances that are sold on the street resemble the gruesome methamphetamine chemically and in their mind-bending allure. (Noncontrolled substances that work quite well for ADD are atomoxetine [Strattera], bupropion [Wellbutrin], and guanfacine.) And, big finale, all of these dopamine-stimulating drugs can cause psychosis, not to mention insomnia, loss of appetite, anxiety, and withdrawal reactions of bludgeoning sedation as the mind struggles to replace the depleted dopamine vesicles. Bottom line, things that increase dopamine generally tend to lead to psychosis, anxiety, insomnia, and poor appetite—one of the cardinal symptoms of depression. (When depression really solidifies, it's not uncommon to see features of what one would define as psychosis including quasi- or frank delusional beliefs.) And of course schizophrenia is marked by a hyper-dopaminergic state.

As the brain expanded under the influence of evolution's steadfast approval, especially in the area we call the granular prefrontal cortex, intentionality and consciousness increased. Decision making was moving beyond the realm of genetics to the individual crosshairs of cerebral interpretation. This process was refined as the brain expanded, bringing in memory of past experiences to help guide decision making as well as simulate future experiences and imagine ourselves in, for example, future conversations. Thus evolution, which rules on genetically determined traits over long expanses of time, was boxed out of decisions about an animal's moment-to-moment, micro-behavioral repertoire. While some behaviors may fall under genetic control, others are learned and respond to influences unique to a particular environment. But since behavior can affect survival and reproduction, old Momma Evolution wanted to stay in the game.

Appointing a deputy, dopamine, evolution washed its hands of the process of ruling on every little behavior a Homo sapiens engaged in. This deputized policeman could issue awards for good behavior, all the stuff evolution sanctified. Yearning for organic matter's timeworn endurance, procreation was huge as was food gathering with hunting and predatory brutality so all got the dopamine dollop.

So evolution enthusiastically approved a system wherein behaviors that increased the likelihood of an animal's survival and reproduction were rewarded, like the old Pavlovian dog, and those behaviors that decreased the likelihood were basically ignored. That deputy, dopamine, an ancient neurotransmitter, when secreted, would maximize the chances that an animal would repeat the most blessed behavior. A behavior that was appetitive, like gratifiers of hunger or procreation, got the dopamine jolt like one's morning cup of coffee. Those behaviors that were detrimental or unlikely to be appetitive were not so rewarded. The brain's expansion into the craggy shore of free will ultimately necessitated a surrogate natural selection system that worked immediately. In essence, the brain shortened the time frame of evolution's decision making process through its persuasive deputy dopamine and did so in a flash of real time.

If for example a mouse pushes a lever in a box and gets a reward of a food pellet, it then gets a dopamine jolt. After a few repetitions, pushing the lever itself is rewarded with dopamine. For whatever reason, dopamine was a reliable behavioral odds increaser. At some point it became associated with pleasurable sensations, but it was not necessarily these sensations that influenced the behavior. It was dopamine itself.

"Ho, hum, so what?" you're saying. Here's what. *Imagine that dopamine is implicated in some of the most devastating diseases known to man like schizophrenia, Parkinson's disease, Huntington's chorea, Tourette's syndrome, and many others like ADD, Alzheimer's, restless legs, stuttering, substance abuse, and so forth. When you drink your morning coffee, shoot your daily dose of meth, or take your ADD meds, you are stimulating dope-uh-meen!* The euphoria of exercise (along with endogenous opioids) comes from dopamine. (By the way, exercise helps Parkinson's patients and caffeine intake helps prevent it.) It is also known as a love chemical, part of the reward and pleasure of sexual activity.

But the dopamine that served Mr. Neanderthal well may not serve us quite as well as we grow into maturity and become contemplative, decision making, conceptual beings far from the physical and experiential grunts we used to be in the jungle. Therefore dopamine may not be the most suitable or necessary in our current situation. Most likely our caveman had a strong dopaminergic tone, a tone that helped him survive. But now other neurotransmitters (like acetylcholine) presumably balance that tone, especially in the motor systems, and that dopaminergic tone hits the proverbial brakes as we reach maturity. This happened evolutionarily as man became increasingly more verbal and contemplative. So, my theory goes, as we reach maturity dopamine is suppressed allowing for a clear brain theater for the pirouette of conscious thought. The ego, which is the sixth sensory organ, delivers these concepts, syntactically organized, to the forebrain and gains strength from this dopamine suppression. This becomes the standard operating procedure of the mind. Anything that interferes with this modern blueprint is detrimental and results in mental or physical illness. Dopamine suppression is involved in gating out unwanted stuff from our brain theater. It promulgates our new 2.0 modern mental complexity and is a fulcrum for crisp, logical expression. All of these not-unremarkable benefits synchronize in an orchestrated cabal as we grow up. It did so as we evolved from wordless grunts to chatty jabberers.

A new view of the dopaminergic saga now emerges in which the old salt takes a turn of events, a crass demotion that places it under the reins of the modern mind. Taking its queue from the (nigrostriatal) motor system where old man dopamine is brought into karmic balance with acetylcholine to hyphenate the smooth coordination of movement without which jerky clumsiness punctuates our reach, with language's onset we now could suppress dopamine in the central tracts (mesolimbic, mesocortical) of the brain's bowels as well. We learn to do this as we race through adolescence so that by age fifteen or so the verdict is in and dopamine settles into its new demotion, and in bowing to this hogtie certain illnesses show a tendency to abate with maturity such as ADD, Tourette's, and stuttering. Thus my theory, simple in its ability to solve numerous problems in the essence of mental illnesses' oblique landscape, bolsters a decided link between language and dopamine binding. And the far side of that composition is that when dopamine un-suppresses, the core of mental illness is unearthed. Mental illnesses display a brutish loss of this dopaminergic restriction, schizophrenia being the ultimate example, and the talented scientific mind of man has designed numerous and devious chemical restraints on this unfettered regression. To further muddy already murky waters, all of these dopaminergic backslides fall into the wily charms of entropy that seeks to deliver us back to a state of lower energy and less organization with its ultimate goal a return to that inorganic universe from which we recklessly emerged. So we are putting together a picture of mental dysfunction that relies on a reversal of a reversal, an undoing of dopamine's tortured fate, all under the guise of entropy's persistent criminality. It's my view that entropy and gravity, that labor twenty-four/seven on our unified entirety, play a greater role in illness than has as yet been our pleasure to acknowledge.

This downgrade or suppression of dopaminergic tone during adolescence re-creates what happened as man became a contemplative animal. Different brain chemicals suited this new configuration, and it seems obvious to me (and many others) that what worked in the jungle, desert, African plains, and so forth didn't suit the modern latte lover's intellectual abilities. Decision making overruled any Pavlovian chemical reinforcer. Mentality 2.0's new sedentary, contemplative embarkation required neurochemical dexterity never before required by Mr. Neanderthal. Socialization demanded greater mental complexity as well, a sort of social strategy impulse control and a whole new set of skills that our cave dweller didn't, nay couldn't, with his 1.0 mind, possess. As survival of the fittest gave way to survival of the most reproductive, new talents met the demands of procreative success. The brash stimulation of dopaminergic vigor was not as essential nor was its crude reinforcing. A shiny new temperament emerged as neurotransmitters other than dopamine came into prominence aided by sexual selection and the increased likelihood of promoting one's gene pool with memes or cultural fads rather than mutations. While organicity progresses by accidents (mutations favored by entropy), suddenly gene pool articulation proceeded by a desirability factor, a need to be chosen, not just exerted in procreative ardor. Cultural norms took precedence in finding a mate over physical fitness. In short our relationship with dopamine had to change and that cantankerous submission transformed the human landscape. This led to advances in braininess but opened up the potential for problems in those for whom our new dopaminergic dalliance wasn't quite as secure or even possible. *Perhaps what separated sapiens most from their brute Neanderthal and Denisovan and Naledi brethren was this ability to suppress dopamine under language's symphonic spell.*

And so what does all this (dopa) mean? Would it surprise you that every single chemical that is used to treat schizophrenia blocks dopamine receptors? One chemical called reserpine used early on depletes dopamine stores, thus having the same effect. Why the need to block dopamine from connecting to its receptors, the keys that fit neatly into the lock that triggers the next nerve cell to fire? That excessive Neanderthal dopaminergic tone has mysteriously reappeared in individuals in their late teens or early twenties, and when it does it explodes with a vengeance. In 1 percent of the population, the feisty dopaminergic tone that was gradually suppressed during maturation suddenly makes a dramatic (or in some cases a slow and inexorable) reappearance like a vaudeville act that was booed off the stage but is now back for an encore with its crappy jokes and corny routines. When this happens we call these people schizophrenics. They have been recaptured by the primitive, entropic, childlike state of mind before language. Fortunately dopamine-blocking agents work fairly effectively for most and restore gating, but they are not without ugly side effects, and unfortunately, some individuals do not respond as well as others nor can we, in our less-than-infinite wisdom, predict who will and who won't.

Like the former friend who slept with your girlfriend, our amicable relationship with dopamine took a hairpin turn, a momentous renegotiation. Dopamine got fired and was suppressed, replaced by the monster of contemplation with its gorilla propensity for problem solving, meaning dopamine now had to be tamed. Too bad, dopa, since your Pavlovian assignment was now replaced by a younger, more talented and cheaper coworker: deductive reasoning. Evolution, in its matronly wisdom, came to the inevitable conclusion that thinking was so good, dopamine had no need to carry on. While we still get a surge of dopamine after sex or appetitive enhancement, the prefrontal cortex now receives input with far more credibility than a swift puff of dopamine. Thinking transformed the landscape, iced age-old predators, and paved the earth in our own geodome. In an apoplectic fit, evolution reveled as we now adopted a paradigm of dopamine suppression. One could go so far as to say that *dopamine suppression is one of the greatest achievements of a Homo sapiens* that those other species like the Denisovans, Neanderthals, and Naledi could only dream of. Once language birthed contemplation and reason, we literally conquered the world.

But a downside surfaced as we realized that this new friendship dealt some collateral damage. Multiple illnesses now surfaced from this new dopaminergic detente, a role that fostered ego strength; allowed for thinking itself by an energetic gating out unwanted intrusions to the mind space; expanded the brain with chemicals like BDNF; facilitated our adult mental complexity 2.0; and, as if that weren't enough, crucially defied entropy. Schizophrenia reverses this suppressive process and reinstates the primitive organization big time, the one before language, in its entirety. Waiting patiently until the brain has reached a certain level of biological maturity and a heavy load of repressive fervor including the Freudian contents, it roars back with insidious intensity. And it is ruthless in its pursuit of entropy's lost kingdom, taking as its mission deflation of the ego (which got us out of Dodge, the 1.0 mentation, in the first place), demoralizing the sufferer. Dopamine de-suppression is exquisitely painful and dissects away all modalities of gratification, leaving the sufferer in a state of anhedonic impoverishment and primitive thought process. The schizophrenic's negative symptoms reveal the most entropic state of all, akin to sleep, and the sufferer lacks motivation, enjoyment, logical sequencing, affect, awareness, enthusiasm. They've reached the entropic trough, satisfying entropy's fondest desires (although return to inorganicity in suicidal behavior is its fondest wish, greatly elevated in schizophrenics) and reestablishing its lost kingdom once again.

As previously mentioned there are different tracts of dopamine. Specifically there is the nigrostriatal, which has to do with movement; the mesocortical; the mesolimbic; and the tuberoinfundibular. They assume different roles, different passions in the dopaminergic theater. The tuberoinfundibular has to do with the release of prolactin, a hormone used during pregnancy to promote lactation and breast swelling. When we block dopamine with schizophrenia drugs we sometimes see breast swelling even in men, as well as sexual dysfuction, if the prolactin level surges. But it doesn't always follow that footpath. In the nigrostriatal tract there is a balance of dopamine with another neurotransmitter, acetylcholine. In fact we first learned to suppress dopamine in this motor system. Dopamine and acetylcholine have to remain relatively equal, partners in a coordinated duet. If we block dopamine excessively, acetylcholine will surge and cause disturbing motor symptoms like restlessness, acute muscle spasms, tongue thickness, rolling of the eyeballs, and other undesirable phenomena. We can treat these by then toning down acetylcholine with what are simply called *anticholinergics*, meds that suppress dopamine's partner, acetylcholine.

Thus with a dopaminergic tone that's insufficient, we may get things like restless leg syndrome, which tends to be more common in ADD. If the dopaminergic tone is too high, we may see symptoms such as tics or involuntary movements as the acetylcholine rages to burst its way through the overbalanced dopamine. Another such manifestation is stuttering. We can treat these with dopamine-blocking medications that punch down the dopamine. Parkinson's disease is a gradual, late-onset deficit of dopamine as the cells that produce it deteriorate, almost a mirror image of schizophrenia, and Alzheimer's disease reflects a lack of acetylcholine.

If we had to put in order the conditions that result from dopamine dysregulation from maximal to less intense it might look like this.

schizophrenia

mania

depression (with poor concentration)

anxiety disorders

obsessive compulsive disorder

nigrostriatal motor disorders like Parkinson's, Huntington's chorea, Tourette's, and restless leg syndrome, stuttering

The motor disorders are manifested in the nigrostriatal tract and reflect an imbalance between dopamine and acetylcholine. If dopamine is paltry, then acetylcholine will surge and induce a state of restlessness. We see this in ADD and restless leg syndrome, both treatable with dopamine inducers. If dopamine is overbearing, acetylcholine will have to punch its way through like a prizefighter with jerky outbursts as in Tourette's or stuttering, both treatable with dopamine-blocking agents. Obsessive compulsive disorder (OCD) is potentially an in-between state with both motor and cognitive features. Classified as an anxiety disorder, OCD has motor features since the patient is driven to perform physical acts (compulsions) in response to obsessive ideation. In its sister disorder, Tourette's syndrome, there is no thought component, just an involuntary motor tic that bursts through a weighty dopamine blanket. For example, the obsessive compulsive may get a recurring idea that if he doesn't check his door lock repeatedly on the way out, someone will break into the house and steal everything he owns. He thus is driven to perform the motor act of door checking repeatedly by an irrational thought that has no purpose but to facilitate that act. Here an ideation greases the wheels of a physical action. We treat OCD with high-dose serotonergic medications like Prozac. Indirectly we are damping down dopamine. Huntington's chorea is another dopamine-related disorder responding to antipsychotics that block, putty-like, dopamine receptors. Parkinson's disease is the opposite, the ultimate in dopamine deficiency.

The little bugger dopamine has thrown a chemical wrench in the system, which in contemplative essence, moves us to starlit heights and leaves a trail of bodies in its wake. The crest of our renegotiation with the pleasure chemical left us with an incomplete treaty, and dopamine doesn't always keep its word. Yet with our stellar science we've been able to manufacture chemical congeners that redress some of these imbalances, though imperfectly, a tribute to our enhanced brain power.

I would assert that *all mental illnesses reflect a disturbance in the reception of conceptual stimuli (thoughts) in the forebrain supplied by the ego, a response to dopamine de-suppression. As a corollary, all treatments for mental illnesses restore the predominance of the conceptual forebrain input by damping down dopamine.* This dopamine theory has been the predominant theory of schizophrenia for the past five decades or so.

This deprivation, as I've mentioned, happens in sleep when the forebrain is deprived of sensation in the dark, muffled, slumber milieu. It goes into sleep mode and begins self-stimulating with dreams. In schizophrenia the same situation exists: the primitive organization before language reasserts itself and the forebrain interprets this sudden vacuum of usual input as sleep and thus, logically, begins to self-stimulate with hallucinations in an effort to avoid atrophy, a common occurrence in schizophrenics. Perhaps the predominance of auditory over visual hallucinations in schizophrenia reflects the closeness of words to concepts. Dreams, on the other hand are largely visual. The sleeping brain is cut off from all visual stimuli but not so completely from sound. The waking schizophrenic has full visual input, eyes wide open, and is therefore swarmed by auditory hallucinations that are not as frequently visual.

Stimulants that push boluses of dopamine are known to cause a psychosis indistinguishable from paranoid schizophrenia. They are re-creating the dopamine surge that happens when the primitive organization that existed before language resurfaces in schizophrenics, a chemical simulation that vanishes once the offending agent leaves the bloodstream although eventually it can obtain permanence depending on the duration of use. Psychiatrists have all seen this.

And what of Freud in this biochemical landscape? How can the couch play a role in the field of chemical battlegrounds magically transformed by language and evolution? The answer revolves about the ego-enhancing nature of psychoanalysis, which fertilizes the dopaminergic suppression. Not only that, but in freeing up repression's coffins buried in the unconscious, more dopamine is released for the suppressive mechanism, making therapy and psychoanalysis enhancements in the dopaminergic waters. The interpretations of the therapist and analyst in undoing repression free up more dopamine and move the ball down the field to reinforce dopamine's binding, which strengthens the gating process and promotes mental complexity against the wishes of entropy. Contemplation, of course, solidifies dopaminergic suppression, releasing BDNF and VEGF into the mind's garden for enrichment. What one might venture to label mental health involves both dopamine suppression and a healthy conceptualizing apparatus in the executively powerful dorsolateral prefrontal cortex area, along with strong gating. Mental illness specializes in dopamine de-suppression and interference with the thinking paradigm that was born of language's illusive gifts. Therefore, dopamine suppression, with its gating function, may well be the neurochemical equivalent of Freud's concept of repression.

Part II

Chapter 4

Johnny's Not the Same

In October I began to fracture, but I did not recognize it as fracturing, and I was told so many things that month, but I was not told that I was losing my mind again.

—Esme Weijun Wang, *The Collected Schizophrenias*

Is that really you? A cruel joke is played on some schizophrenia sufferers, as if the enveloping madness weren't enough, and now they have to confront the theater of their loved ones or support system being replaced by doubles! Oh, yes, not a problem, Mom is just some lady who happens to look like her, sent in by Acme Acting for a little clownish impersonation. That wouldn't disturb you, would it? These two-bit actors living there, taking up precious space in the house, and grunting in indelicate omission of your mom's true self? It kind of looks like her, but well, the rest is an awkward deviation from her core-ness.

Some schizophrenics (a small percentage) have something called *Capgras syndrome*. (A syndrome is more of a collection of symptoms.) In this unusual phenomenon the patient believes that his parents (or some other important figure) have been replaced by gruff impostors, doubles. These schizophrenics are convinced that the designated key individuals are the spitting image of their parent, let's say, but just ain't them. They may know with startling conviction that their therapist or psychiatrist or spouse has been replaced. And the replacement is not a kindly doppelganger but rather a sinister look-alike sent there to persecute the patient and has no earthly connection to them at all. Of course if that person for example asks the schizophrenic to take their medication, the schizophrenic, being skeptical of this impostor, will likely refuse knowing that they are being scammed into doing something injurious. A scene out of a horror movie or a *Twilight Zone* episode, this can be very frightening to the patient. Imagine what it'd be like if people you didn't know were living in your home and pretending to have your best interests at heart, imitating your loved ones. You might want to flee or report them to the police or, rarely, try to harm them. (All of these things have happened.) It certainly severs the bonds of even the longest-standing relationships. It also demonstrates that for the schizophrenic the ties that bind are tenuous at best. It serves the purpose of the patient's paranoia and treatment resistance and, of course, furthers dopamine de-suppression. "Why should I follow any of their advice?" the schizophrenic thinks.

The person may project all sorts of evil impulses onto the double. (When I say project, I mean attribute to others an impulse that really lies within ourselves but is denied conscious awareness. Remember that consciousness thing when we become aware of something as long as it is connected to its word label? Another way to make something conscious is to send the impulses through the sensory channels. Dreams are sent to the visual and auditory channels to reach consciousness.) I should note that most symptoms serve multiple purposes. This represents an economy of function, killing multiple birds with one stone.

The reason I bring this up is that for some families it is almost the exact opposite of what happens to them. Their child, a nice kid, maybe a little shy, somewhat self-absorbed, is living a fairly normal existence. They could be average students or whiz kids, social butterflies or wallflowers, nerds or art freaks. Whatever the case, they show only modest if any evidence of a problem. (Research continues into children who might be called *ultra high risk for psychosis kids*. This is an attempt to identify who is going to become schizophrenic over the ensuing years with as much certainty as possible. If we can reach a high level of certainty, then we can make a decision about some form of intervention, medical or otherwise, to try to prevent this transformation.) They may have cognitive deficits of a mild type. Research shows that relatives of schizophrenics are prone to more cognitive deficits, thinking oddities one might say, than others. Exactly what this means is hard to define.

Then suddenly everything changes. It may be a dramatic change, or it may be slow and inexorable but subtle at first. It may not happen until the child goes off to college. It may begin in high school, the later teen years. But suddenly, *Johnny is not the same*. He's more isolated and he's doing things that are atypical for him. For example, he becomes obsessed with a particular movement or cult-like group, conspiracy theories, religious preoccupations. He adopts a mission, like saving the starving children of Africa, or promoting a specific video game. He may become obsessed with a particular person he sees as a savior or the greatest living being. He may talk about nothing else or stop talking much at all. Something is going on inside his mind but he has a hard time conveying this to others and often does not want to. (I'm using the masculine pronoun for no particular reason as there are about as many female schizophrenics as male although males tend to have a worse course and slightly earlier onset.) He may start obeying what he sees as the rules of this mission. Probing his inner thoughts is often met with resistance.

Things inevitably progress even further as there may be a loss of personal hygiene and slovenly dress may develop as concern for personal appearance wanes. Withdrawal from society at large is not uncommon, and his statements become illogical, which may be subtle at first but is obvious later. Rationales for what he's doing are often nonsensical, and caregivers become frustrated and concerned. Attempts to get help may run the gamut from emergency room visits to homeopathic remedies to enrolling Johnny in a military academy. Professionals will hesitate to make the diagnosis, which will sentence Johnny to a lifetime of difficulty and lower expectations. However, it's best to bring that person to a mental health professional ASAP. Let that health professional get a sense of the situation and attempt to establish a connection with the patient. Unfortunately, we know it takes much too long for schizophrenics to be diagnosed and to be treated. There is potential loss of future functioning and brain atrophy in that delay when professionals err on the side of underdiagnosis or family members on the side of denial.

There may be dramatic changes in behavior: a sudden impulsive overdose, wild rants and misplaced euphorias, bursts of anger followed by remorse. The schizophrenic may completely renounce his parents since, after all, they are the bearers of reality and they may insist on his receiving some form of help. The primitive organization, in its ruthless quest to take over the mind, uses ideas to promote its cause. For example, like Bethany Yeiser in her book *The Mind Estranged,* a schizophrenic might say, "I am the second coming of Mother Theresa and need to live on the streets in poverty to help the starving people of Africa." The schizophrenic often wants none of his parents' realism. He knows where the truth lies and will defend his right to believe in it. There may be a whole host of paranoid statements tantamount to accusations against well-meaning people whom he labels persecutors, or he may passively go along with treatment while often lacking in conviction. Suddenly he is saying that poison gas is emanating from the vents, ladybugs are talking to him, or the moon is looking directly at him. Suicide attempts may occur as the patient realizes that he and reality are not on the same page and that something powerful, unstoppable, and nefarious is trying to take over his mind.

This takes us back to the first paragraph. Johnny has suddenly become someone different. He is not gone, not deceased, yet he's not himself. One might call this Capgras syndrome in reverse. The parents have their son yet this person is not him, he is someone else, and that someone is devastatingly altered.

Not surprisingly the family wants their relative back. (Schizophrenics with the best prognosis have an actively concerned family. Many are not so lucky. Others find their family giving up on them over the lengthy course of the disease. It's a very painful situation for all involved.)

* * *

This is where our journey begins, the same place a schizophrenic's journey begins, the start of the illness. Everything you've read leading up to this was preparatory work. Like a hunter going into the savanna to hunt lions, we needed to gear up, don our camouflage jacket, strap on a rifle, and check the magazine. Without all that we'd be less likely to nab the big game, those nasty preternatural forces that lay mental illness at our feet and expose the inner fleshy pinkness of their essence. Yes, we are going big game hunting and shall unleash the mighty force of insight against the stubborn megalith of insanity.

> *During this time I had several dreams to which I did not then attribute any particular significance, and which I would even today disregard as the proverb says, "Dreams are mere shadows," had my experience in the meantime not made me think of the possibility at least of their being connected with the contact which had been made with me by divine nerves. I dreamt several times that my former nervous illness had returned; naturally I was as unhappy about this in the dream, as I felt happy on waking that it had only been a dream. Furthermore one morning while still in bed (whether still half asleep or already awake I cannot remember), I had a feeling which, thinking about it later when fully awake, struck me as highly peculiar. It was the idea that it really must be rather pleasant to be a woman succumbing to intercourse. This idea was so foreign to my whole nature that I may say I would have rejected it with indignation if fully awake; from what I have experienced since I cannot exclude the possibility that some external influences were at work to implant this idea in me (Schreber 1988).*

Thus begins one of the most famous and best-documented journeys through psychosis of all time. The writer, Daniel Paul Schreber, went on to spend many years in a German asylum in the late 1800s and early 1900s. There were no medications available at the time (Thorazine, the first antipsychotic medication, was not used until the fifties). Electroconvulsive therapy (ECT) was not developed until the 1930s in Italy. Schreber's erudite documentation should be read by anyone interested in schizophrenia as it's a very intricate and masterful work by a devoted thinker. Someone you've heard of did read it and found it quite interesting. His name was Sigmund Freud, and he set about using Schreber's memoir as a case study. He never met Schreber but subjected the memoir to psychoanalytic investigation resulting in his classic explanation of paranoia. (Schreber was quite paranoid and believed, among other things, that his doctor, Flechsig, was out to kill him.)

Schreber had suffered a previous illness that was described as a hypochondria, meaning excessive worry about disease, accompanied by weight loss. He was hospitalized for this in an asylum and cured by none other than Dr. Flechsig. Schreber's wife was so grateful for this that she kept a picture of Dr. Flechsig in their home. There is a phenomenon that I call the *on-the-way-to-psychosis reaction*, which means, essentially, that someone who is getting an inkling that their mind is about to be taken over by a powerful force reacts to this knowledge with an illness. It is the stuff of science fiction weaving its way into real life. Imagine if you felt a gradual, inexorable conviction that something insidious you had no understanding of, something Thor-like in its power, was threatening to take over your mind from within. This awareness might be presented to you in various ways like a religious vision or the fleeting appearance of a deceased loved one. You might find yourself behaving at times very differently or you might notice a change in your thinking. A suspiciousness of everything or of certain things or a belief that you must save the world, for example, might come on very strongly yet you would ignore these aberrations or envision them peripherally with a skeptical indifference, trying to neatly fold them out of your mind.

As it continues, the situation might induce periods of anxiety or depression that wash over you in salty waves. There might be withdrawal and depression leading you to burn yourself or stop eating, to avoid everyone and give up on things you used to worship. Perhaps you get angry at people you love and become defiant, erratic, or oppositional. You might seek out recreational drugs in a desperate attempt to rectify the situation, to do something when no one else is offering any real solution. (But street drugs are simply gas on a fire. Not one mental illness is improved by street drugs or alcohol. They are invariably made worse by them, and the drugs may create illness where there was none while leveraging the rapidity of its onset.) Probably the best analogy to a schizophrenic experiencing the onset of their illness is someone who has just swallowed a hallucinogen like a bitter cup of ayahuasca stew and is waiting skittishly for its effects to take hold.

One schizophrenic young woman I treated years ago had successfully completed college. (Details of all patient examples are altered to protect anonymity.) She was working in a loan review area and became convinced that she was being observed and followed by loan inspectors. This progressed to a conviction that the FBI was tailing her. She was brought to the hospital by her mother who had noticed changes in her behavior like odd statements and suspiciousness. When I met with her she was minimally concerned by her recent changes in thinking; however, to please her mother she had agreed to come into the hospital. (Insight is a major prognostic indicator. The greater awareness that one is ill, the more likely one is to improve. Schizophrenics are known to stubbornly deny their illness to the point of delusional absurdity. Even the brightest patients convince themselves they no longer need medication, and when for the tenth time, they crash back into florid psychosis. They still maintain that outpost of denial [Saks 2007].) This patient, however, agreed to take medication—Risperidone if I remember correctly. She quickly started to get better. When I went home for the weekend I thought she was going to be much improved by Monday. But by Monday she'd stopped taking her medication.

As if Alice could be in that wonderland but still be oblivious to the change in locale, the trait of self-observation that may be strained in the best of us dissolves completely into dust for many schizophrenics who tend to be uncritical of the warp in their demeanor. Some can hang onto a self-observing eye that struggles to inform, clinging to autonoetic awareness. It's the six-million-year-old entity that ate your brain, and you stumbled zombie like into the fog. The primitive organization has you in its grip with the alacrity of a starving parasite. This is no innocent bystander but an avaricious mental ninja that will fight for every inch of your soul. There is no turning back, no polite reversal, no buyer's remorse: the octopus has got you with its slimy tentacles and has no intention of letting go but rather digging in, and then the sea beast, in its seductive charm, manages to persuade you that anyone attempting rescue is your enemy.

One Sunday morning at church, I met a young Caucasian woman who had just returned from a year-long stay in Africa. After returning from Africa, she took time to debrief with a family from church and to discuss the "poorest of the poor" she had interacted with in Africa. Friends told me she was an unusual and special person. Her name was Susan. When I met her for the first time I saw Susan had a wisdom that seemed quite unlike that of any person I had ever met . . . I became jealous of her . . . I began to see her as one of the best people on earth and barely human. I was becoming delusional. This was the first time I had ever been delusional (Yeiser 2014).

Yeiser's observing mind, while intact, was unable to stop the inexorable takeover by the primitive organization. Yet a part of her knew, standing aside in drone-like view, that things were changing for her and not for the better. Yet when budding schizophrenics come to us for answers, we've had none to give them until now. The psychonaut in some jungle outpost in Peru at least knows why his mind is starting to feel and behave weirdly . . . He's just downed some hideous Peruvian vine potion or smoked the excretion of some tortured toad. The new schizophrenic has no such explanation available to them.

When The Doctor comes in, he brings backup—another attendant, this one not so nice, with no interest in cajoling me or allowing me to keep my nail. And once he's pried it from my fingers, I'm done for. Seconds later, The Doctor and his whole team of ER goons swoop down, grab me, lift me high out of the chair, and slam me down on a nearby bed with such force I see stars. Then they bind both my legs and both my arms to the metal bed with thick leather straps . . . Moments later, I'm choking and gagging on some kind of bitter liquid that I try to lock my teeth against but cannot. They make me swallow it. They make me.

Someone watching me. Something watching me. It's been waiting for this moment for so many years, taunting me, sending me previews of what will happen. Always before I've been able to fight back, to push it until it recedes—not totally, but mostly, until it resembles

> *nothing more than a malicious little speck off to the corner of my eye, camped near the edge of my peripheral vision.*
>
> *But now, with my arms and legs pinioned to a metal bed, my consciousness collapsing into a puddle, and no one paying attention to the alarms I've been trying to raise, there is finally nothing further to be done. Nothing I can do. There will be raging fires, and hundreds, maybe thousands of people lying dead in the streets. And it will all—all of it—be my fault (Saks 2007).*

At some point, some people get the sensation that something is trying to take over their mind. (Here we see that Elyn Saks, at least, was able to recognize this force and collapse it. She was in control of it . . . up until then. Finally, inevitably, control is wrested from the individual and the primitive organization conquers.) Not just a vague sensation, it is a movement so powerful as to be unstoppable. This is not a friendly coup but an avaricious revolt, an insidious occupation. Yes, it is the stuff of science fiction movies. It is a slimy tentacle from the inside out. Slow and inexorable for some, it comes on like a freight train for others. There will be efforts to sidestep this bull, with red capes and tortuous dives to the ground. But inevitably it comes back to haunt them. Why they have been so blessed we don't know. The search for a schizophrenia gene continues to no avail. Like most major illnesses it tends to run in families but not always. It makes its appearance in the late teens to early twenties but there are exceptions. Just as we presume it to be genetic, there is just as much evidence that it is entirely random. (This statement will be considered controversial in the medical community.) Identical twins have exactly the same genetic makeup. They are genetic carbon copies. If you have an identical twin who happens to have schizophrenia, you have a 40 to 50 percent chance of also having schizophrenia. Why not 100 percent? If two exact clones can diverge in regard to having an illness, it clearly is something more than genetic in origin. Yet we keep looking. At some point we must admit that there is a certain randomness in the incidence of this plague. The central paradox of schizophrenia is that it still exists at all. Schizophrenics have a much lower rate of reproduction (fecundity ratio) than normals and a significantly impaired ability to survive and succeed, which means, according to some guy named Darwin, that they should quickly become extinct. They have not, do not.

Other theorized causes of schizophrenia include a virus (patients born in winter months have a slightly higher incidence), birth trauma or infection, schizophrenogenic mothering, preternatural possession by the devil, and prenatal infection or trauma. The only theory that has not been promoted is "None of the above." Evolution is not listed as a primary potential cause and the nature versus nurture debate rages on. Certainly both could potentially be involved as genes and the environment interact. There are risk factors that may increase the odds that you will get the disease such as urban living. It's hard for people to accept that possibility. Yet for some illnesses there is a degree of randomness that governs the incidence. In short, schizophrenia could happen to anybody, including you, but if you are over twenty-five consider yourself basically out of the woods as it is rare after that age.

Decreased functioning is a hallmark of schizophrenia although there are notable exceptions. Patients and families are told they simply have to set their sights lower. Many schizophrenics cannot work. They live in group homes that cater to chronic mentally ill patients who are disabled. Some work low-stress, low-demand jobs that provide minimal income and even less challenge. Hopes of a career, of a family, of educational attainment often end. The schizophrenic is still there but no longer themselves.

What has happened? This puzzle is as ancient as the Oedipal myth. Where did this invasion originate? What heinous thing has eaten the patient's brain? Schizophrenia was once called *dementia praecox*, an early dementia. Yet most would concede that it is not a dementia since schizophrenics know the date, can perform basic functions of daily living, and so on, while some are as high functioning as any Wall Street executive. What is it? Would it surprise you to hear that this illness has yet to be clearly defined after centuries of debate? This book does exactly that, defines a centuries-old conundrum, separates it definitively from other mental illnesses, and answers the central paradox and the anthropo-parity principle that questions why schizophrenia exists as 1 percent of populations worldwide.

And what of the doctors who treat this mystery? We sit in offices or on hospital wards and see these individuals we call schizophrenic. If the diagnosis is not clear initially, it generally becomes obvious over time. We look for the dramatic symptoms (what we call *positive symptoms)*: the hallucinations, the bizarre behavior, the delusions. We look for the less-dramatic symptoms (what we call *negative symptoms*): lack of facial expression (*affect*), lack of interest (*anhedonia*), illogicality of speech. We notice a downturn in functioning, a difficulty relating to others, an occasional dishevelment, and the lack of social mores that goes with it. We take note of age of onset and changes in demeanor reported by the family. Eventually the diagnosis simmers to a clear broth.

So we prescribe medications to sufferers of this mysterious malady, but treating a malady one doesn't understand breeds resistance and promotes stigma. Sometimes the medications work well, other times barely at all. We know just so much about how they work (every single one blocks dopamine receptors in the brain while the newer atypicals block serotonin *and* dopamine in a ratio greater than one). Aside from that, we don't know much about these medications except that they tend to have lots of side effects and that patients (who often don't agree with our diagnosis) go off them regularly. We definitely know that the patients are better off on the medications than off them; tons of research tells us that. For patients with a lot of symptoms we often see a dramatic reduction in those symptoms over a few days. That is not a cure but it does help the patient regain contact with reality as their mind clarifies and the demons slink away. In short, we're treating an illness we don't understand with medications we only partly understand and receiving mixed results for patients who don't believe they're ill (*anosognosia*). It's time we understood this illness, and mental diseases as a whole, much better. To do that requires context, imagination, and a willingness to discard one theory for a better theory if that's what the facts demand. Come with me down this tortuous road.

Chapter 5

New Rules

> *One patient remarked, concerning some "red bricks," that these bricks "are my transformed thoughts on love, nothing but red love." Another patient took the black color of doors as signifying "dying." A third patient saw in the twisted legs of a table the meaning that "the whole world is twisted." In sum, through the extreme shrinkage of distance between the self and the surrounding world, the objects and states of affairs in the schizophrenic's environment become completely unstable.*
>
> — Heinz Werner and Bernard Kaplan, *Symbol Formation*, pp. 256–57

When I asked one of my female patients, "What's the difference between a rhinoceros and a hippopotamus?" her answer was "The rhinoceros must be the male." Clearly she was using the fact that the rhinoceros has a horn as proof of masculinity. Another female patient told me she had cured herself of Alzheimer's. Another patient told me he was dead. When I asked him how he was able to stand there and talk to me he had no answer. He persisted in this delusion (called Cotard's syndrome) for quite a while. Another patient believed that food was being poisoned and was barely eating at all. He said that if he ate, other people would suffer, and hundreds would die. These forms of communications obviously play by different rules than the communications of normal people. They display illogicality of various types and use rules that don't apply to normal speech. Just as with children, the distance between self and others (objects) is not fully established. What they need, in a sense, is more distance from the words they choose (Werner and Kaplan 1963). They use words differently than most people use them, and their words can become objects themselves with meanings more idiosyncratic than universal, syncretic with dual meanings, objects of gratification more than communication.

An ill-defined infirmity, schizophrenia puts the brain tissue in jeopardy while the mind struggles to regain its balance in a new land. Treatment requires urgency to prevent frontal brain deterioration, the atrophy of disuse. A once modern brain 2.0 is in regression, and the replacement model 1.0 is not rational. Though not a dementia, there is reversion to the childish ways of thinking from infancy—puzzling to the adult world as this sudden replacement module 1.0 sputters in its effort to redress the calamity. The once furious and focused mind collapses, victim of an infirmity so elaborate and sinister as to baffle the keenest philosophers and confound the theologians.

So if schizophrenia is not a dementia, what is it? It's my (and many others') belief that a core symptom of schizophrenia is cognitive regression. Their thinking plays by different rules, and clearly something has happened to their reality testing and their ability to prioritize incoming stimuli. Elyn Saks described her psychotic decompensation as everyone at the Super Bowl screaming at her at once (2007). Anyone would become easily overwhelmed by this bombardment. Remember the ego's function as a word processor, sixth sensory organ presenting concepts to the prefrontal cortex? It's as if the ego has been bowled over by some force and left crippled, barely sputtering along. The ego is also charged with prioritizing incoming stimuli, and with it dopamine suppression has a powerful gating function. In schizophrenia it's as if the ego took a huge hit and now the individual is limping along without one. So dopamine is de-suppressed, gating dissolves, and the mind floods. Others have commented on this ego-leaching phenomenon (Tausk and Feigenbaum 1992). Something knocks the ego out of commission and triggers a return to primitive, prelinguistic organization.

The replacement is one of childlike thinking, poor reality testing, delusions, hallucinations. It's as if an operating system 1.0 from the eighties was installed in a new MacBook Pro designed for the modern adult 2.0 upgrade. Some functions might work, others will not, and some will work in part. And so it is with schizophrenia as some ancient operating system has replaced the adult brain 2.0 that schizophrenics developed over the first twenty years or so of their life. They are fully adept at language and they have achieved the rules of adult thinking. If their IQ has been tested, it's within the normal spectrum for most or might be way above normal. So it's not a question of low intelligence or dementia nor is it possession by the devil or drug ingestion although some schizophrenics do use drugs.

> *In dealing with the schizophrenic's handling of language, it does not suffice to determine merely that the patient's utterances deviate in certain external characteristics from the utterances of normals. One must at least consider the possibility that the basic attitude of the patient towards language differs from that of the normal adult. One cannot conclude from the fact that the schizophrenic may exploit conventional linguistic forms in his utterances that he actually uses or regards language in the same way as does the normal. Indeed it is our contention—following from our theoretical viewpoint—that the extreme change in orientation towards the world that occurs in schizophrenic states must entail a fundamental transformation in the patient's attitude toward language (Werner and Kaplan 1963).*

My response to this is, "Of course it does!" The schizophrenic has regressed to a time before language! The primitive organization extant in our primitive relatives for over six million years or so now recaptures the unwitting schizophrenic soul. So how is it that they can speak at all, you ask? Because they've spent the first twenty years of their lives learning language, and their ego has used language to enhance their cognitive functioning. What happens when the schizophrenic process sets in is that they are now functioning with the six-million-year-old 1.0 brain that man had before language. Their thought processes have returned to, perhaps, that of a five-year-old child yet there are aspects of the adult mind 2.0 left untouched by the tsunami of schizophrenic invasion. In the developmental schema of things (Werner 1948), they are less specific, more diffuse, less organized, and less hierarchically differentiated in speech. "In sum, autistic regression leads to a collapse of distance between vehicles and referents (objects)" (Werner and Kaplan 1963).

You remember that it is the ego's job to distinguish between self and others. But when the schizophrenic process sets in, the ego takes a hit. It goes back to the time when there was a minimal, nonverbal ego and the powerful Captain Kirk of modern language didn't infuse humans. It is the ego that gets us out of Dodge and pulls us away from the 1.0 primitive organization of pre-language humans in the first place. Thus the ego becomes enemy number one of this primitive organization. What ensues is a conflict that is part and parcel of this psychosis or loss of reality. The schizophrenic process, or what I call the *primitive organization*, like an evil coup attacks the ego mercilessly. Its goal? Total domination of the brain. In particular it wants control of the voluntary musculature. One can make a solid comparison between OCD (obsessive compulsive disorder) and schizophrenia in this regard, and they are often seen together. OCD is also a first cousin to Tourette's syndrome. In the latter a dopamine excess forces another neurotransmitter, acetylcholine, to surge forward, resulting in motor tics. Dopamine-blocking agents can tone down the dopamine, thus leveling the balance between it and acetylcholine. This is a movement disorder without cognitive influence. In OCD, as we've seen, an obsessive idea influences behavior, and the the patient might believe she must check the garage door ten times or she will be robbed. The acetylcholine urge is led to gratification by a (false) concept. In that sense OCD is a movement disorder with delusional cognitive influence still gratifying the acetylcholinergic urge. In schizophrenia the primitive organization also seeks to gain control of voluntary musculature. It does so with relentless persuasion, commanding derogatory voices and belief systems that play into its hand; in short, the influence of ideation is much greater since the person has gone back to primitive thinking without the benefit of dopamine suppression and its gating function. Their world is internally bound and not externally focused, and external reality serves more as a stimulus to their inner thoughts than as a bearer of outside evidence. Tourette's is more likely located in the nigrostriatal tract, which is a dopaminergic motor bundle in balance with acetylcholine. The compulsive movements of OCD patients are nigrostriatal but more under the influence of mesocortical or mesolimbic winds. Thus with OCD and schizophrenia, control of voluntary musculature becomes the illnesses' greatest goal, enlisting delusional ideas in their quest along with dopamine de-suppression.

The primitive organization is often very punitive, an insidious power broker. There is always a sense of degradation by the primitive organization, the process being regressive and asocial. What I call a *social strategy* is not possible in the regressed mentation schizophrenics find themselves in that, after all, is a return to a time before civilization. Pre-verbal humans existed in a time before the superego was created for civilization's benefit. In short, the individual has been recaptured by the six-million-year-old pre-verbal brain of our ancestors yet is fully familiar with words. The brutality, desperation, and incivility of the caveman is reexperienced in psychosis. (Autistic patients have usually not attained this level of development. They have not allowed language to drag them out of Dodge in the first place, so to speak.)

Remember the difference between primary and secondary process thinking? Primary process thinking is an echo of daydreaming, an autopilot type of mental rambling that is uncontrolled and nonparticipatory. Imagine our poor Neanderthal sitting on a rock staring out over the canyon and contemplating how he will survive the day yet he has no words. His brain has expanded to help him thanks to the benefit of some mutational changes millions of years earlier that expanded the prefrontal lobes, silently steering him toward success. Sitting there on that cold rock, he feels hunger, he knows danger, but he has few tools to help him. Finally he gets up and starts his day with a nagging dread inside him, the fear of predation. What will he run into today, a snake, a flood, a bear? What gnarly contortions will he have to go through to satisfy his and his family's needs? Will he stumble on to see another day?

This is the state of mind that the schizophrenic returns to, insidious, tenacious, and terrifying. What does all this have to do with primary and secondary process thinking you ask? Just about everything. Secondary process, deliberate, rational thinking suppresses dopamine, strengthening the ego and a gating function that yields a pointed focus on the details of the external world and revels in realistic thinking while primary process thinking de-suppresses dopamine and returns to the primacy of internal mental constructs. We are no longer in that primitive state of mind. We have an expectation of survival and have conquered most of the obvious predators that plagued us day to day with our superlative contemplation skills born of language. We sit in our comfy heated houses while the snow flutters like butterflies around us and we bask in a feeling of general contentment. Freud said that the most we can expect in life is to be reasonably happy. Clearly life is not without stress, but generally we don't have to leave the house spear in hand anticipating becoming something's lunch (nor did all hominins even have spears).

It was language, the power-boosted ego, that talented word processing, and the sixth sense that learned to deliver concepts to our brains, prefrontal lobes most likely, that pulled us out of the prehistoric primitive organization 1.0. How? Words made their debut about fifty thousand years ago, and we've seen that words can be used in many ways. (Perhaps it was the result of an ever engorging sense of intentionality in animals that pointed fate's fickle arrow to language that then brought us, with a horse's leap, over the fence.) Children use words as proto-symbols, something that antedates symbols, that are laden with idiosyncrasies. Words are, in a sense, imbedded in the individual child's encapsulation early on and in schizophrenics as we have seen, being much more focused on their inner mental experience than on external reality and communication. Initially words are not sharply differentiated for the child and carry with them a baggage of idiosyncratic meanings and memories and feelings, the funky inner themes that influence what the child says. Their communications are not so much universal as autistic, individual in meaning and expressing things other than what a pure, clean, dissected symbol should. The latter tap dances into view once there is gating of those swirling inner preoccupations, which cannot happen without dopamine suppression, a gradual accretion of adolescent's enrichment. When a child says, for example, "bathy bath," he or she is not just referring to soaking in a warm tub to get clean. They are referring to many things: the fuzzy warm feeling that they get, the glee of having Mommy help them suds up, the scariness of going under water, and so on. All these inner notions are expressed simply with those tiny words. Early on words are syncretic, having multiple functions, and then gradually these autistic meanings melt away and the words become universal symbols as dopamine is suppressed and the inner is gated out in favor of the carnival of external reality. By age ten, when he says "bath," he means cleaning up in a warm tub. His words have been un-imbedded from his own swirling inner themes; they are communal, focused, understood by all. This process occurs as we grow up, and it occurred over the past fifty thousand years in us Homo sapiens. Neanderthals could only dream of getting the glittering treasures that language gave us. They had no facility for sound production with their smaller voice boxes and maybe they could not suppress dopamine much. And finally, this process, this symphonic movement, defies entropy as it lands our brains in a richly organized and energized Xanadu, the human brain being the most anti-entropic (or as I refer to it, antropic) substance in the universe.

So what, you say? *This process delivered man to a new frontier!* We went from childlike thinking to adult thinking and along with it found greater differentiation of ourselves from words and, incidentally, from others. This evolution, perish the word, delivered us to, yes, secondary process thinking 2.0 with dopamine suppression and enhanced ego function, gating out the swirling surround sound of childhood's mentation. For the first time, humans could participate in their own thought processes! The ego learned to deliver concepts to our brains for its contemplation, which furthered the pedestal of dopamine suppression and released brain fertilizing chemicals like BDNF and VEGF. Our thought processes are arrow pointed, well organized, and focused on problem solving in the external playground. We can construct theories from points of data as the ego scoops out items from the preconscious fish barrel that gratifies the whims of the subject we are contemplating, decorating the thought swirls with precious words and delivering them to us via auditory sensory channels. But we don't stop there. When we think or study or read our brains expand, secreting nourishing chemicals so that our nerve cells root, the dendrites branch out, and we grow mentally from this activity. Conceptualization is not only a way to contemplate but also a way for us to expand our brains and our egos, perhaps the real motive behind thinking from its inception, and all of this in defiance of grumpy entropy's mafia-like desires.

It is known that congenitally blind people are immune to schizophrenia. How can our theory explain this? Congenitally blind people, as opposed to those who become blind later in life, acquire language without the benefit of any sight template or hint of common-day outlines. They have had no visual sensory experience, and as such, when they learn language this becomes their sight so to speak. The conceptual reality it provides replaces the behemoth sensory input from the eyes, and therefore any insidious takeover of their conceptualizing ways would be defended against with the vigor of an ogre. The process of language acquisition, ego intensification, and dopamine suppression is massively intensified, constructing a conceptual reality that fills in sight's massive vacuum. This is not true when language is acquired by the sighted. Vision accounts for the majority of sensory input. The blind also need conceptualization of their day-to-day outings in a dark universe, all of which implies that there may be, in extreme circumstances, the ability to fend off the primitive organization's kung fu barrage but only in this rare condition.

We learn to use our brains in this 2.0 iteration as we grow to maturity, and it becomes the modus operandi of the adult's mental theater. In order to do this we learn to suppress dopamine as we did in the nigrostriatal motor tract to improve coordination. (We borrow this talent from the motor area.) Language has helped organize and complexify the mind much to poor entropy's chagrin. We use our ego to supply concepts to our forebrain, but when the primitive organization resurfaces (surging dopamine), the brain is sensorily vapid as it is at night and acts as if it is asleep. (In sleep we return to the prehistoric, primitive organization 1.0 of our ancestors, reexperiencing the entropic nadir of our bumbling past.) It begins to self-stimulate like a severed, twitching muscle with dreams that, for the waking schizophrenic, are experienced as hallucinations. Every mental illness interferes with the practiced chorus of our new 2.0 operating system, that is, the ego spoonfeeding concepts to the forebrain, or dorsolateral prefrontal cortex. They do so by de-suppressing dopamine, and this is assisted by faulty thinking in the form of delusions that are a fulcrum back toward a primitive, internally focused state of mind. Every chemical treatment we have for mental illness restores this basic blueprint, reestablishing the supremacy of the forebrain's prefrontal cortex (that blossomed like a cabbage rose over the past three million years or so under the various mutations that expanded it like NOTCH2) and the gracious reception of thought forms.

Skipping along, our new partner in sanity, language, took us by the hand and marched us out of the Mad Hatter's living room, and we found ourselves basking newly in our own thought processes for the first time. We replaced intuition with solid, deductive reasoning. And not surprisingly, we would soon conquer the world, defeat every foe, and orchestrate a new universe of institutions, golden rules, and enlightening commandments. The birth of religion, science, education, art, jurisprudence, and other institutions followed civilization's lead.

The potential tsunami of dopaminergic de-suppression, as we've noted, portends a plethora of mental and some physical problems. While the new conceptual mind with its stronger ego brought us out of the Dodge of primitivity, it left us vulnerable to slippage as well as imbalances between dopamine and acetylcholine as we forged a new compact with this key brain player. A new treaty was signed, but as with most treaties we quibbled on the details and still do. When we treat major depression, for example, we tone down dopamine indirectly by turning on the autoreceptor thermostats using neurotransmitter enhancing medications called *antidepressants*. When we treat mania or schizophrenia we directly suppress dopamine with antipsychotic medication that blocks the receptors that receive it like blue silly putty in the holes of a sieve. We treat Tourette's, Huntington's chorea, and stuttering similarly. Those schizophrenics who use their minds conceptually have a better prognosis, and there is a whole movement of cognitive restructuring to try and reimagine the schizophrenic brain toward 2.0 functioning. There are dopamine deficiency states as well; for example, restless leg syndrome, ADD, and Parkinson's disease, which is practically the opposite of schizophrenia. All of this is set in place by the process of language acquisition and ego muscularity and dopamine's new suppression. We each go through this suppression as we mature.

Which is why the primitive organization wants none of it. The ego has pulled us away from that antique 1.0 mind, and in order for it to even consider reconquest of the territory of our psyches, it must bludgeon the ego. A very insidious confrontation ensues, resulting in psychotic symptoms that exploit every opportunity to denigrate the ego while voices urge the sufferer to hurt themselves and make negative statements about them and their capacities. There is a return to what I call *encapsulation* . . . the internal preoccupation of children as opposed to the externality of the real world. They are encapsulated in their own minds and are, like Mr. Neanderthal, experiential and physical beings. And of course there is cognitive regression to more childlike rules of thinking.

Within this playground the psychosis unfolds, and for someone like Schreber, it can be intense. He felt as if the guards in the hospital were going to drown him and believed bees were sent to attack him. He talked to birds and saw two suns, yet despite all this he managed to emerge relatively intact by exercising his mentality toward explanation of his mental theater. And his ego had to accept the delusion that he had been turned into a woman, after which the intensity diminished—a bargain entered into with the primitive organization. (We also bargain with the devilish primitive organization via certain symptoms like delusions or various preschizophrenic diagnoses that serve the function of sacrificial lambs to forestall a worse, schizophrenic outcome. One could say that all other diagnoses are attempts to ward off the granddaddy of them all, entropy's delight, schizophrenia.) Schreber kept his mind intact by extensive thinking and writing a lengthy, literate memoir. In the end, he proclaimed loudly that God had failed in making him demented.

* * *

Schneiderian symptoms were named after the German psychiatrist who elucidated them, Kurt Schneider. They are considered by some to be pathognomonic (meaning highly typical) of schizophrenia, and when we uncover these gems we feel more certain we've clinched the diagnosis. The list includes auditory hallucinations, particularly those that echo the subject's thoughts out loud, for example, two voices blabbing about the subject in the third person or discussing their thoughts. Also included are experiences of having thoughts inserted into our mind by an external agency in a telepathic zapping, thoughts being broadcast somewhat like tweets out to others, the removal of thoughts, and passivity in which our movements or lack thereof are both under the control of some evil external force (Tausk and Feigenbaum 1992), and delusions arising out of normal perceptions (Tikka et al. 2016). These symptoms can be explored directly by the doctor (although patients won't always admit to them) and are somewhat common in the regressed schizophrenic. Schneiderian symptoms are correlated with lower levels of BDNF suggestive of a deterioration of brain function. (You'll remember that BDNF is like a fertilizer for the brain. When we use secondary process thinking, this also promotes BDNF.) They also reflect a lower D2:D4 ratio (second digit to fourth digit ratio and smaller in schizophrenics). The latter may be related to lateralization of the central nervous system that, as you recall, is a function of language acquisition, which, as I have noted, assists our commandant the ego in forming the florid conceptual consciousness 2.0 that stands apart from the early, 1.0 primitive organization. Imagine yourself sitting in a restaurant and everything you think is blasted out of a loudspeaker as you think it so you look around and get a powerful inkling that everyone seems to be hearing your mental flotsam and jetsam . . . Scary to say the least.

What I call *primitive organization theory* hypothesizes that in his or her late teens or early twenties, the schizophrenic is swarmed by the tsunami of a primitive organization replacing pieces of their adult mind. This primitive organization is hyper-dopaminergic and cognitively regressed, the 1.0 pre-verbal operating system of our prehistoric brethren. It originates from a time before language when, for a mere six million years, hominins like animals did not communicate in words (Leakey 1994). The onset of language about fifty thousand years ago changed all that and the trajectory of evolution along with it. Homo sapiens, by suppressing dopamine, were able to acquire learning, and conscious thought, ultimately expanding their minds and creating civilizations that arose coincident with the complexity of their thought processes. They were therefore eventually able to neutralize many of the forces of adversity that had interacted with mutations to propel evolutionary protoplasm forward. Once these forces were neutralized and man had developed a cozy expectation of survival, natural selection or survival of the fittest went into free fall and was replaced by survival of the most reproductive. While we are still deep in the midst of this novel course correction, we are not yet fully there, and some individuals do not traverse the transition smoothly enough, making them vulnerable to recapture by the primitive organization 1.0, a process fueled by entropy's longing for disorganization and energy troughs. This reinstates a hyper-dopaminergic tone that can be mollified by antipsychotics. (Pre-verbal cave guy was nothing if not hyper-dopaminergic as it was majorly expressed in the motor nigrostriatal tract and used as a Pavlovian reinforcer for behavior that gratified Momma Evolution's fondest desires as opposed to entropy's.) Unfortunately, higher structures like the burgeoning prefrontal cortex are short-circuited and in a time-dependent manner may fizzle into atrophy. As they do they may self-stimulate, resulting in hallucinations akin to dreaming.

Who were we back in the jungle playgrounds of our Neanderthalish forebears? Just another furry species, we had our own grungy habits and a brain big enough to tip the scales. But like most animals, we were nothing but physicality and sensory intake to funnel our efforts toward appetitive behaviors instinctually driven. Without words or thoughts, we flew on instinct with the help of our autopilot mind. A little dopamine dollop served as a Pavlovian reward when we pleased evolution. The ego, in charge of movement and little more, did its best to steer the ship toward a low bar of success: survival and procreation. But it was a wimpy ego, doing little to differentiate us from the world around us. Words were waiting in the wings and with them thinking, that promise of advancement over the rest of organicity.

Once language arrived about fifty thousand years ago, the ego became a dynamo word processor extraordinaire. The mind organized around language like iron filings sucked to a super magnet and the brain enlarged on one side to accommodate it, which we call *lateralization*. A thin preconscious layer developed in which words magnetized to their intended targets as labels and then a shuffling process of these words into sentences, using mysterious syntactical rules. Captain Kirk developed a finely honed rigamarole in which words were pulled up from that preconscious cubby, organized syntactically into conceptual sentences, spiffed up like an old shoe, and then presented to consciousness through the auditory nerves for inspection, awareness, and contemplation. And that consciousness, in the prefrontal lobes no doubt, held those concepts up against a template of truth, gauging their veracity in a court-like drama. The glittery new verbal ego could be considered another sensory organ whose products were not sounds or images but actual thoughts. This new conceptual input enriched the brain bouillabaisse by promoting chemicals like BDNF and VEGF that increased dendritic expansion and also had a mood-elevating effect. This was gradually automated so that the first steps were done whiz bang, unconsciously.

A new verbal, conceptual consciousness arose alongside the experiential consciousness of Neanderthal Nate. Concepts could now be inspected in the spotlight of awareness with that golden template. In order for this to succeed, dopamine had to be suppressed, which then gated out unwanted intrusions. This developmental process is repeated by each individual as they grow up. The child progresses from a diffuse, poorly integrated cognitive being overly bound by internal inputs to a more integrated, hierarchically structured, and differentiated mental system pointed squarely toward the outside world, giving birth to consciousness. Along with conceptualization, the brain function level of humans marched toward adult thinking 2.0. These adult rules of thinking conformed more to reality than the child's inner preconceptions. The ego boundary between self and other was hardened as the ego became empowered, a beefy Captain Kirk. Homo sapiens now had an adult cognitive edifice, a conceptual consciousness with which to evaluate ideas and subject them to reality testing templates and syntax, gating out unwanted intrusions, all of which demanded and relied upon the suppression of our tap-dancing friend dopamine not just in the motor tract (nigrostriatal) but now in the central tracts as well (mesolimbic and mesocortical)—something not possible until there was language.

The suppression of dopamine is complete between age 12 and 25 with a sharp division between conscious and preconscious (the Freudian metapsychology). Something Freud called an *Oedipus complex* has to be mastered, theoretically, by each individual, which also empowers our friend the ego swelling the gating balloon. Each individual, according to Freud, comes to terms with parental desires and masters them. These developmental processes carry the preschizophrenic to an adult mental wooden framework just before the onset of their illness in their late teens or early twenties and preceding the brain's endpoint of physical maturation, which also includes, according to my theory, the suppression of dopamine. Once that happens it may be too late for the primitive organization to try to gain possession of the mind. This implies a window of opportunity for the primitive organization after which it can no longer launch its attack, dopamine being in some way under lock and key.

Somewhere in that time frame, as the central nervous system is tinkering with maturity, a primitive organization 1.0 from our prehistoric past reasserts itself in schizophrenics. The evolutionary yellow brick road that drags us from childlike thinking to adult cognitive function 2.0 is usurped by the mindset that existed for some six million years before language. That does not mean that schizophrenics cannot speak or think but that their conceptual framework 2.0, focused on the external reality, has been replaced by the primitive framework 1.0 of the past and of children dwelling in the clouds of their inner themes and conceptual tributaries. It's as if a modern computer's operating system 2.0 was replaced with a primitive operating system from the eighties. Many basic functions would work but on a much less sophisticated level, and other functions would be absent entirely. Such is the mindset of the invaded schizophrenic. This simplicity and lower level of complexity and energy (since the energy used for gating is released) is a euphoric blessing to entropy that wants nothing more than for substances to exist in its laws, the laws of inorganic matter, and it finds the de-suppression of dopamine the salve to its dreams.

Thus a nearly mature Homo sapiens suddenly gets infused with the brain of our Neanderthal ancestors, a primitive, ancient mind. As if science fiction had to prove how real its horrors could be, there it is, a relic occupying the space of a modern mental edifice, leaving the sufferer to reinhabit the jungle outpost of our prehistoric brethren. "Talk to me," the schizophrenic says. "I come to you from ancient times." They did not know they were going to be on a mission from the past, reanimated, with caveman-like equipment. It just happens, and it's up to us to learn from these disconnected souls about who we were, who we are, and the crux of the difference.

Getting back to Schneiderian symptoms, the mind of the schizophrenic is usurped by a primitive simplistic mentation millions of years old complete with dopamine de-suppression. However, the adult conceptual erector set that processes words into concepts is still there as evinced by the patient's ability to speak (and think!), but their brain space is often flooded with ungated stimuli rushing in like a swarm of hungry bees. The adult ego continues to create some concepts although not on the order it used to, but the primitive (pre-verbal) ego landing strip 1.0 experiences these words in a way comparable to an infant lying in a crib hearing its parents jabbering. Words are objectified and bounce off this primitive pre-verbal receiving post like ping-pong balls. The degree of distance between the word-producing, conceptual ego and the receiving experiential consciousness is perhaps the most salient parameter in regard to the appearance of Schneiderian symptoms. The sensation of thinking one's own thoughts is replaced by a sensation that thoughts come from some distant foreign shore. They no longer flow naturally and joyously from the conceptual generator doodad of the brain through the auditory pathway to conscious awareness. (One can assume that anything that makes its way to our consciousness flows through the sensory pathways. Dreams flow largely through visual pathways to our sleeping consciousness. Another option might be through the memory pathways. I've seen patients whose primary symptoms were distorted memories.) One might call them *ego dystonic*, meaning that they don't rest comfortably in repose in that ancient prelinguistic ego receiver. Words may feel inserted like a pin in a ripe mango as they are unintegrated or experienced as echoes or as coming from a distant loud speaker or another person. Hallucinogen users also report such experiences.

Paranoid mechanisms come into play with quirky delusions when the feeling of word insertion morphs into a sensation of bodily embedding instead with spacey proclamations of microchips having been surgically implanted in the brain or teeny radios buzzing a vapid static in one's ear canal. These are fairly common delusions in schizophrenics and not uncommon in strung out methamphetamine users. These delusions of inserted objects instead of words may be experienced as robotically influencing the person's movements, generating crackling sounds or an endlessly annoying static. What we call the *influencing machine*, or a sinister, octopus-like device with grasping tentacles, according to Freud (Tausk and Feigenbaum 1992) is symbolic of a penis. Thus the implanted microchip is unconsciously an inserted penis although not in the usual receptacle but displaced to one's gray matter. I might note the choice of inserted object or influencing machine has progressed from the crude box with batteries of Tausk's patient to the contemporary lore of microchips, persecutors hacking into one's computer or cell phone, and so forth. The phallic metaphor has progressed to digital-era symbols of mystical power. Even delusions modernize with the times.

Words, now object-like and experienced as oddball intrusions, or dystonic, may become these phallic penetrations that can be cannonballed outward to others or broadcast forth as a sort of rejection of them. The schizophrenic then looks around the crowd searching for any telltale facial expression that clinches the conviction that others are hearing those thoughts and invariably comes up with someone even in blandly indifferent faces. This indemnifies in their minds the experience of broadcasting or mind reading by others. The conviction that someone else's thoughts are somehow in one's own head is another form of disorienting mental concatenation and again correlates with the phenomena of word un-integration since the thoughts are being ground out by the old Captain Kirk, who has been demoted and replaced by Captain Primitive. Kirk continues to bellow forth stern orders as if he were still calling the shots but Captain Primitive doesn't recognize these orders and dismisses them as utterly irrelevant.

Tausk and Feigenbaum (1992) also highlighted the ego weakness of schizophrenics. As we've seen, this is a result of replacement of the adult, concept-driven Kirk of our 2.0 operating system by the more regressed, weaker ego of the primitive organization, the remnant of our prehistoric past that we have yet to leave behind, like a spaceship that's still in the earth's gravitational persuasion. This Captain Primitive has fuzzy childish boundaries between self and world, self and others. Tausk and Feigenbaum (1992), referring to "loss of ego boundaries," states, "This symptom is the complaint that everyone knows the patient's thoughts, that his thoughts are not enclosed in his own head, but are spread throughout the world and occur simultaneously in the heads of all persons. The patient seems no longer to realize that he is a separate psychical entity, an ego with individual boundaries." This is actually an expected phase of early childhood. One way a child learns that his thoughts are not in others' brains is when the child realizes he or she is able to lie. In the schizophrenic mentality, this makes it seem like odd-looking words are bouncing from their brain to others' psyches with impunity. It is coupled with a caveman-type operating system of thought 1.0 typical of perhaps a five-year-old child, and with a psychedelic sleep-like encapsulation, setting the stage for a hatter's table of psychotic phenomenon. All of this is in the context of dopamine de-suppression, poor gating, and entropic entreaty. "Come back, come back," entropy implores, "all is forgiven." Feminine urges play a role in which control by an outside entity may be a wished-for gratification and is an outlandishly distorted submissiveness. Schreber (1988) experienced many such delusions including that his eye movements were controlled by "little men." Ultimately he was convinced that he had been anatomically transformed into a woman.

Once the primitive organization 1.0 storms the Bastille, the newly minted 2.0 house of cards crumbles like a sandcastle in a tornado. Since we dance back to the entropically sanctioned 1.0 state of mind in the dreamy land of slumber nightly anyway, the executive centers of the brain interpret this grotesque turn of events as a waking state of sleep. But it's a sleep that's stubbornly persistent, and in order to avoid the deterioration of being so summarily deprived, the brain stimulates itself, as it does nightly, with waking dreams we call *hallucinations*. The schizophrenic can still summon up their exquisitely honed contemplation skills, but they're cramped to say the least by thorny, unguided intrusions that cantilever the brain space with penguin-like irrelevancies since the gating function of dopamine suppression is lost. Saks (2007) described this as everyone at the Super Bowl screaming at her at once. However these manufactured Broadway dream productions may not be enough to forestall deterioration of the prefrontal cortex, aching in its need for conceptual input to avoid dissolution, an atrophy routinely intensified with each new psychotic relapse, which is why we need to treat schizophrenics early and with a full-court press.

This primitive organization acts like a two-bit dictator in a crumbling country. When voices that comment about the individual chime in, reminiscent of a child in the crib hearing its parents' praise, it's all denigration and sinister commands. This new mental cabal needs to target the force that put it out of power in the first place, our kindly ego Captain Kirk. Delighting in ruthlessness, the productions that emanate from frontal lobe deprivation are all commandeered into pejoratives like a cache of rebels who take over the government's artillery. Sinister voices urge the schizophrenic to harm themselves or scream insults designed to demoralize and intimidate. (In fact the voices that urge the schizophrenic to commit suicide give direct expression to entropy's desires, the voluntary return to inorganicity, evolution's greatest failure.) It's a ruthless takeover with no mercy shown to the individual once the process lifts off like a hot air balloon. At its ultimate victory it has complete control of the individual who is now its marionette, the strings pulled by an avaricious puppet master who is not satisfied with anything but an iron fist of control. This, of course, all curtsies to the raging bear of entropy that sees schizophrenic regression as its ultimate victory in the quest for a scattering disorganization and sinking energy.

Echoes, bouncing words like beach balls, a windy mumble of barely heard voices, these all occupy the schizophrenic ungated mind space. Like a playground of irritable kids, it dances to its own tune, tinkering with logic and ultimately rejecting it. A random word, tossed over the net by the modern ego, bounces back, tearing off to the side and rolling away. The ego tosses the ball up again and slaps it over the net where it bounces indecorously off the other pre-verbal ego, which barely notices its intrusion. The game is lost, another victory for entropy's desire.

It is time to redefine the word "psychosis." That vague word, which is roughly equivalent to "craziness," can be punched up in more specific detail. Psychosis is a product of cognitive regression, a march back to the prehistory of our pre-verbal caveman past. The individual, thinking with more primitive, childlike rules, says crazy things just as children do. This is a byproduct of their altered relationship with words and the use of different assumptions, childish themes unencumbered by external reality with internal thought streams dominating the underlying iceberg of their verbal productions. This happens once the primitive organization storms the Bastille. Just as kids say the most amazing things, schizophrenics reveal the change in their rules of thought in their verbal utterances and learn to say as little as possible, disconcerted by the reaction of normal adults. If they spy a table with twisted legs and say, "You see, the world is twisted," they get sanctimonious admonishments. If, on seeing a black door, they remark that it signifies death, they get harried looks.

But one could say that "psychosis" is a general term also seen in drug ingestion and other states of compromised brain function. I would assert that the mechanism is the same: usurpation of the modern mind with the primitive organization under dopamine de-suppression, which resides in us all, with all of its logic bending ramifications. Don't waste your money on an ayahuasca enlightenment journey to Peru. All you're doing is allowing your primitive mind to take over while you're awake. You can let it do so in sleep quite nicely and without the puking and insect-ridden dangers of Amazonian encounters.

Chapter 6

Let's Talk Medication, Paul Schreber, Prepulse Inhibition, Charles Bonnet Syndrome

But in the first years of my stay at Sonnenstein the miracles were of such a threatening nature that I thought I had to fear almost incessantly for my life, my health or my reason.

—D. P. Schreber, *Memoirs of My Nervous Illness*

Back to our lonely psychiatrist, sitting in an office somewhere, while patients march in and out. The patients are suffering and seem to have lost some ability to function. They say strange things and are often eerily illogical. Again we distinguish between positive symptoms and negative symptoms, the latter settling in like a creeping fog over time. The positive are the more flashy ones: voices, visions, delusions, bizarre behavior. These tend to occur early on in the course of the illness when our primitive organization, fueled by entropy, is railing in battle against the ego, its enemy number one. The negative symptoms include alogia, or illogicality; constricted affect (blunt facial expression); anhedonia (inability to enjoy things); poor motivation; and a certain degree of ambivalence about basic life issues. That's not to imply that positive symptoms cannot exist as the schizophrenic gets longer into the illness but that they tend to shift, coast, and be less dramatic. Paul Schreber, who was placed in an asylum around 1900 and whose memoirs Freud used to construct his classic explanation of paranoia, had symptoms that ultimately lost their vitality. Despite being quietly delusional he was deemed by a court to be dischargeable.

So what does the poor psychiatrist do? Thorazine, the first antipsychotic (the class of medications we use for schizophrenia and also known as neuroleptics because they can affect the nervous system) was developed in the fifties. What did psychiatrists have before that? Two options: lobotomy and ECT, electro-convulsive therapy. The gruesome lobotomy was fairly barbaric with a needle shoved into the patient's forebrain and swooshed around, liquifying frontal lobe tissues. This was not a treatment most patients wanted, and their families were horrified. Yet it did often work, ratcheting down the agitation, creating a sense of comfort in the patient, a denominator of relatableness that was perhaps rather childlike. My theory is that it ended the conflict between the primitive organization and the executive functioning centers of the mind. But whatever the mechanism, it was permanent, sometimes caused seizures or even death, and was a rather gruesome option. Clearly there was some desperation in this treatment.

In the 1930s ECT made a dramatically welcome appearance. Invented in Italy by Cerletti and Bini, it became a very popular alternative to the lobotomy for an obvious reason: it didn't liquify any brain cells. It does usher forth a grand mal seizure just like what an epileptic experiences when they unexpectedly drop to the ground flailing their limbs in supercharged muscular contraction. The whole procedure lasts about ten, fifteen minutes and is much more refined today with the patient asleep and his muscles in mega-paralysis. The machines used today deliver less electrical wallop and still manage to extract a full grand mal seizure. We can adjust the pulse width and the frequency of the brain zap, and we use square wave instead of sine wave electrical forms, which is a more efficient delivery system. It's called a *modified seizure* in that the patient is knocked out by anesthesia and is paralyzed so there is supremely attenuated shaking and muscle contraction. The usual stolid course is three times weekly for a total of six to twenty treatments, and then we either call it a day or put the patient on a maintenance regimen that is once a week or less. They can come in from home if they have a reliable person to drive them, stay with them, and hover around them for several hours once they get home to allow the anesthesia to metabolize out of their system.

Initially this mind-bending treatment was significantly cruder than it is today. Instead of anesthesia they gave the patient a slight dizzying knock-out shock and then the full wallop to induce the grand mal. Several staff strong-armed the patient's limbs while the patient went into wrenching, whole-body seizure activity that occasionally imparted spinal fractures from the intense conniption. The machines used were cruder, delivering a behemoth shock in a less-refined way. As a result ECT got almost as dismal a reputation as the lobotomy and then in the seventies was dealt an ogre's blow when the movie *One Flew over the Cuckoo's Nest* burst on the scene. The film portrayed ECT as a mind-numbing, behavior-bending punishment designed to silence obstreperous upstarts in an era of hippy rebellion against authority. It's a perfect example of art affecting a medical treatment.

ECT subsequently went into decline and was underused for decades. Meanwhile the technique underwent a wholesale mechanical enhancement with brief anesthesia soon de rigueur along with paralyzing agents like succinylcholine to minimize fractures. The electrical energy involved slithered down to less-robust doses with excellent results thereby sparing the individual the blockbusting brain crush of the shock delivered previously.

ECT has been coming back into standard use. If one looks on the Internet, there are lots of scare stories and misinformation, but if one scrutinizes the scientific literature, ECT is one of the safest treatments in all of medicine. It is also one of the most effective as some 75 to 85 percent of the thorniest treatment-resistant patients will walk into the promised land with this electrical wonder. So it remains an option, often of dead last resort, and one that most patients eschew over a pill, yet it's part of the psychiatrist's armamentarium and we're glad we have it.

In the fifties, the dark ages, there were what one would call *asylums*, which means "places of refuge." These ghoulish, Halloween-like places were jam-packed with schizophrenics, many of whom were reticent recipients of ECT because there was an echoing dearth of ministrations other than, as I said, lobotomy. So when Thorazine burst on the scene, it was truly the God potion, or the penicillin of psychiatry. Thousands of schizophrenics got better, or better enough to be discharged, all of which fostered a shining period of de-institutionalization. However, it was never funded adequately and patients did not get the outpatient treatment they needed, so some of them devolved to homelessness or were sent back to the dreaded asylums from whence they came.

No one ever expected a pill would be able to reduce schizophrenic's symptoms, and Thorazine, developed by Delay and Deniker, was intended to be a presurgical mellower without zonking the patient to sleep. It worked supremely well so some wise individual decided to try it on agitated schizophrenics, and lo and behold, not only did it entomb them in calm, but it took away or at least pleasantly diluted their raging psychosis. This fortuitous turn of events led pill-inventing pharmaceutical companies to, as they always do, make copycat medications with the same basic recipe. Many types of standard, or typical, neuroleptics (meaning nerve affecting) were simmered from the witches' cauldron of pharmaceutical science. We can, for simplicity, break them down into high- and low-potency bundles with the low-potency medications like Thorazine and Mellaril dosed in the hundreds of milligrams while the high potency were dosed in the ten to twenty milligram range. These medications (and all neuroleptics including the newer atypicals) have one thing in common: they all block dopamine receptors. Furthermore, the intensity with which they do so (or their "receptor affinity") determines their potency in a simple elucidative formula.

So psychiatrists giddily started using these medications en masse and found that they had some interesting if not unfortunate side effects. To list a few, there is tardive dyskinesia, neuroleptic malignant syndrome, intense muscle spasms, tongue thickening, akathisia, weight gain, and Parkinsonian symptoms. Tardive dyskinesia is a neurological syndromic affliction with disfiguring, crummy, involuntary movements. After many years of treatment, usually at high doses, patients started having involuntary lip smacking, tongue thrusting, hand movements of a writhing type, slow tremor, blinking, and grimacing. Unfortunately it was often gruesomely permanent and stopping the medicine worsened it, at least initially. For some the symptoms would gradually diminish or even disappear, but it was disfiguring and uncomfortable and, as I said, involuntary. It had a relatively low incidence, about 4 percent of patients per treatment year.

Neuroleptic malignant syndrome comes on the stage even less often but is more outright dangerous. It's a type of harrowing allergic reaction to the medications in which the patient has a fever, becomes very rigid, and goes sky high with their creatine phosphokinase levels, reflecting muscle damage from that rigidity. This can lead to death and kidney failure, but stopping the neuroleptic often leads to resolution. We use dopamine promoting congeners, the opposite of blockers, to improve the odds in this medical emergency. We warned patients about these and did our best to minimize them. Medicines that kick the butt of side effects, like Cogentin, help with the extrapyramidal effects and Parkinsonian effects like muscle spasms, stiffness, and tongue thickening by balancing the dopamine-acetylcholine duet. Nothing helped for tardive dyskinesia so we monitored for that, and if we started to detect it, we resorted to medication diminution or outright banishment that was usually a dismal bust: the symptoms came back.

Did the early medicines really work? Absolutely. All the studies showed that there were immediate calming influences as the dopamine receptors were clogged with vital blockers, which restored the sacred balance, intensifying the gating function, and then came an ongoing symphony of diminishing psychosis. Acutely, a patient who was flagrantly disturbed with intense, positive symptoms like echoing voice commands, demonic visions, off-the-wall behavior, nonsensical statements, and rambling agitation would improve within hours of downing the pills. They would adopt a mellifluous demeanor, stop mumbling to themselves, forget their preoccupation that the FBI was making a documentary about them, and usually sleep better and start regaining a foothold in external reality. Someone in that desperate frame of mind is walled into their own head (just as infants are encapsulated in their own mentality) in an inner world that is light years from adult reality for the most part, their rules of logic childlike and primitive. They are capable of acting on delusional thinking and they may be responding to command auditory hallucinations telling them to harm themselves or others. They think by rules that are simplistic, blunt, less focused. All of these things improve sometimes within hours after administration of an antipsychotic Thorazine tab.

So what happened? They were restored to a more normal dopaminergic tone, the result of a finely honed suppression of dopamine as we mature. When the primitive organization of the mind storms the Bastille, reasserting itself, it does so with a super-charged surge of hyper-dopaminergic intensity. The dopamine receptor blockers keep that hyper-dopaminergic tsunami blast from stimulating them. Usually it requires 60–85 percent receptor blockage to furnish a Zen-like calm. So we start with somewhat lower doses and then gradually ratchet the medication up to maximal voltage and hopefully minimal side effects. Patients become more relatable; start to converse in a more reasonable, adult way (their minds clearer since dopamine de-suppression brings with it a loss of gating of primitive intrusive stimuli); demonstrate less-kooky behavior as the delusional thinking dissolves; and are less encapsulated in their own psychotic, moonstruck universe. An earlier medication, reserpine, worked by sucking dopamine output from the brain cells, another way to skin the hyper-dopaminergic cat. (This principle is now being used in new tardive dyskinesia treatments.) Clearly, restoring the dopaminergic balance goes a long way to restoring the schizophrenic's karma calm. An obvious attribute of the resurgence of the primitive organization is a floridly hyper-dopaminergic tone due to dopamine de-suppression like a raging canyon flood, another being cognitive regression to simplistic thinking 1.0, loss of gating, and the interruption of secondary process thinking, all bundled together in the absence of dopamine suppression.

If you have water pouring recklessly out of a faucet into a colander and then into a sink and threatening to flood your kitchen and you need to choke off the flow of that (dopaminergic) water, there are two ways to go about it. The first is to turn the darn faucet handle to reduce the flow. Unfortunately the faucet's ancient and creaky so we don't have a means of strangling the dopaminergic flow. What do we do instead? We grab a gooey, tacky putty (Thorazine) and jam up the holes in the colander. We start plugging them and when we get to about 60–80 percent blockage, the water flow is reduced enough to prevent an overflow. When the primitive organization reasserts itself in schizophrenics it does so with a high flow of dopaminergic water. Another way to picture it is that de-suppression of dopamine dissolves the Berlin wall of gating between mentality 1.0 and adult conceptuality 2.0 while dopamine suppression strengthens it.

Mr. Neanderthal was no doubt hyper-dopaminergic. Dopamine served our rough-and-tumble primitive ancestors well. It helped them survive in a predatory environment, and the evolutionary thumbs down or up known as natural selection was tickled pink with it. Evolution used it as a Pavlovian reward treat, and we have more dopamine receptors than any other animal. Does that mean that primitive, pre-verbal humans were all schizophrenics in their mega-dopamined psyche? Well, one might argue that, but I would make a case against it. Here it is. When humans became verbal, the ego became the word processor par excellence. Humans went from experiential animals without an inkling of thought participation to conceptual animals with the ability for the first time to take part in contemplative endeavors (and thus promote chemicals like BDNF that increase the expansion of nerve cells and their endings, called *arborization*). BDNF improved neuroplasticity, expansion, and growth along with VEGF. For the first time we could participate in our thinking *and* make our brains grow, *and* this had an antidepressant potency as well. (The new wonder antidepressant ketamine, which has multiple drawbacks but works instantly to relieve depression, stimulates production of VEGF.) What Freud called secondary process thinking as opposed to primary process daydreaming, the nonparticipatory ruminating typical of depression, defines this type of active thinking. To sum up, language with the help of the ego changed our modus operandi of mental processing to participatory thinking, allowed expansion of brain areas such as the prefrontal lobes, and was a mood booster, at the same time drawing a line between the gruff Neanderthal mind of our caveman past, gating out unwanted inputs and conjuring up a wisp of mental nirvana. It advanced our cognitive presence from a childlike, internally fixated mind 1.0 to one that was externally centered and adult 2.0 with universal communication. (It also turned the mind into an antropy machine with the ability to dilute entropy with mental exertion. The mind became an entropy buster with the help of language.)

All of this was brand new and (my theory) required a *suppression of dopamine* to erect a kryptonite barrier between the hobbling, ancient 1.0 mind and the spiffy new modern adult model 2.0. Dopamine came into balance with other neurotransmitters. There are several dopaminergic tracts in the brain and one of them, the nigrostriatal, is involved with movement. In this tract, dopamine is lassoed into balance with acetylcholine, another neurotransmitter, as we mature to promulgate refined coordinated movements. It is the imbalance of dopamine with acetylcholine that can lead to movement disorders like Tourette's syndrome, restless legs, stuttering, Parkinson's, and a whole witches' brew of other maladies including the side effects of antipsychotic medications since they're plugging those dopamine holes en masse. Thus techniques for reducing or in some cases augmenting the flow of dopamine are used in a variegated garden of illnesses. Why? Because we had to alter our relationship with dopamine once we acquired language and complexified our brains. Dopamine becomes hog tied and clamped to clear away the brain space for clarity of conceptualization and to gate out the pesky, unwanted remnants of our 1.0 childlike brain (and Freud's Oedipal impulses). This is the mentally healthy modus operandi gold standard of our new latte-guzzling brethren. Originally we learned to suppress dopamine as we gained coordinated movement including walking and running and lifting. Bringing dopamine into balance with acetylcholine, this nigrostriatal tract Zen mantra allowed us, as well as Mr. Neanderthal, smooth, gliding muscle coordination. It wasn't until we got language that we were able to carry out this feat of legerdemain in the other dopamine tracts in the brain unrelated to movement. (Speech is a complex movement involving ideation as well.) It is when dastardly dopamine un-suppresses that mental and some physical illnesses bubble up grotesquely. Evolution's legacy is vastly positive as regards language with a huge thumbs up, but the need to alter dopaminergic function as we learn to use words leaves mankind vulnerable to a host of goblinesque, dopaminergic, collateral damage since we ares still getting used to this new neurochemical arrangement.

Entropy is gratified when the primitive organization reasserts its icy grip since it sees an opportunity unequaled in the history of animal brain protoplasm. Entropy knows that no other species has been able to defy it like the kingly Homo sapiens. Our newly minted brains, antropic mechanisms that crackle in complexity, are no doubt the most antropic substance in the universe. But for the 1 percent not yet evolutionarily on board, when the past reasserts itself, dragging them backward toward ancient times, humble entropy exerts its pull even beyond anything cave guy would have reckoned with. This process of de-suppressing dopamine is entropy's desire, and it digs in with gusto at this unexpected windfall. Slapping the ego around, dragging dopamine to unheard of quarters, it releases the not-insignificant energy required to keep the mindset 1.0 buried in the unconscious, an effort that expands exponentially over time. Isolating the brainy frontal lobes to the point of starvation, birthing hallucinations as in sleep, the schizophrenic mind ultimately achieves the most entropic brain state possible, an utter dopaminergic backflow, the lowest state of mental energy possible in the schizophrenic. It revels in a victory more profound than anything it could have bargained for with Mr. Neanderthal, whose brain was nowhere near an affront. No, schizophrenia was not possible in the caveman, who experienced neither modern mental complexity nor chronic residual schizophrenia.

I believe that schizophrenia is a return to a hyper-dopaminergic state from one of dopamine suppression. Medications reduce the passage of dopamine to its receptors just as Thorazine putty blocks the holes in a colander. Every single medication that reduces psychotic symptoms does this, so again, our ancestors may have been hyper-dopaminergic but were not schizophrenic. Schizophrenia is, therefore, a modern invention reflecting the recapture of the mind by the more primitive 1.0 mindset, leaving the adult cognitive structure decimated. Only since the adoption of language and the changes it brought to evolution and our brains did it come about. I would also assert that happiness and suicidality (entropy's greatest victory) are modern inventions as well.

Example: a patient is brought to a psychiatrist by his family. He is twenty-three and looks a bit disheveled but otherwise all right. His mother says he hasn't been the same recently. He told her that if she didn't make some macaroni and cheese pronto, the world would come to a bleak demise. He wrenched the electrical sockets out of the wall in his room because voices were inside blaring imploringly from them, and he wandered around the community at night necessitating a call to police. When found, his explanation was that he was "inspecting the perimeter." He purchased an imposing, eight-inch hunting knife "just in case." He has been eating nothing but macaroni and cheese and hardly drinking anything. He notes that the "precarious world condition" is such that most foods are contaminated and often deadly. His sleep is sporadic and he has abruptly stopped showing up for his job as a cashier. He uses "a little" marijuana "here and there." He looks supremely uncomfortable in his chair in the office and doesn't want to be the focus of attention, seeing no problem with what's been going on. His mother's sister was diagnosed with paranoid schizophrenia years ago, and his mother wonders if perhaps her son is similarly afflicted.

The psychiatrist is already thinking this given the patient's age and his symptoms, which are fairly cookie-cutter for paranoid schizophrenia. He recommends hospitalization for "first break" schizophrenia given the knife purchase, but the patient refuses, declaring that he is not suicidal but that if the world keeps going like this he might become so. He agrees to give the knife to his mother. The doctor urges him to stop using marijuana, and he reluctantly acquiesces, unable to see his use as having any major deteriorative downside. The physician orders tests such as a CAT scan of his brain and a urine toxicology screen and then prescribes Haldol, five milligrams twice per day. The patient is reluctant to take any medication as it might have been contaminated by "world events" and the "industrial pharmacological complex." However, with insistence from his mother who wisely refuses to take him home unless he's reliably saturated with medication, he agrees to take it. The doctor schedules him for a two-week follow-up and tells them to call should any problems arise.

Two weeks later they return and things had gotten considerably better. He was making fewer bizarre statements, would no longer eat only macaroni and cheese, was sleeping well, and had returned to work. His mother was quite pleased up until three days ago when to her utter chagrin he stopped taking the Haldol. He explains this by saying that the medication is probably tainted by the Chinese and that it is part of the "geopolitical complex" and is no doubt affecting his vital organs. Not only that, but his illness has vanished and he's completely fine because he's not hearing the howling tones from the sockets. Of greater concern is that two days ago he went to Walmart to purchase a gun but was fortunately unsuccessful. He explains the need for a gun only by saying that the "situation in Mexico, Chile, and Iran could deteriorate fast." The doctor hospitalizes him against his will, concerned about the potential dangerousness of the patient's deranged weapon fixation. The CAT scan was normal; the drug test showed marijuana only.

So what happens next? Clozapine (or Clozaril, which is the brand name). What I know of the clozapine story is this. It was first used in Europe where they found it to be an excellent medication but unfortunately, rarely, people died from it. They did so because their white blood cell count suddenly dropped and they could not fight infections, and although this tragedy was less than one in one thousand, it certainly was enough for clozapine to be put on ice. Besides this *neutropenia* as we call it (drop in white count of the blood), clozapine had a truckload of side effects. Here are a few in no particular order: weight gain big time and appetite increase leading to insulin resistance and possibly type 2 diabetes; drooling; seizures; constipation to the point of bowel obstruction; and dangerous interactions with other medications like Prozac, which can double the clozapine level overnight. I'm leaving out a lot.

So, you're thinking, who the hell would ever want to expose themselves to this most problematic of treatments? Better question, why would the FDA approve, around 1990, clozapine for use in this country? Because, *it was just that good.* I mean, it was platinum fine. Patients who had been on standard medications for years and doing poorly suddenly did better on clozapine. This was a red-hot shocker. It is well known that the standard neuroleptics lose traction as time goes on. Early on when the patient has a lot of those positive symptoms, the neuroleptics generally work well. As time progresses, they tend to fizzle into a bland dilution. We try medication juggling, additive combos, and even some unorthodox desperation moves such as adding a mood stabilizer like Depakote. More recently, medications like minocycline, an antibiotic, or anti-inflammatorics have been tried with dribblingly flat results. "Why would this loss of traction be?" you ask.

Let us not forget that the onset of the primitive organization's takeover short-circuits those executive brain grids like the frontal and prefrontal lobes. The primitive organization returns the schizophrenic to a pre-verbal state with more childish, primitive, and downright psychotic rules of thinking and much less secondary process thinking. Not surprisingly there is deterioration of those higher structures, which are starving for input and thus self-stimulating with hallucinations. This is why one part of a schizophrenic's prognosis may lie in the extent to which they exercise their cognitive brain power in activities like studying, reading, and contemplation. The ventricles, which contain cerebrospinal fluid, widen to fill the space as the brain tissue shrinks, but medication cannot reactivate tissue that has degenerated. The longer the schizophrenia has had its tentacular grip on the brain, and the greater the number of relapses, which throw more salt on the intracranial wound, the more brain tissue deteriorates. One of medication's jobs is to reanimate like Mickey Mouse in *Fantasia* the short-circuited frontal brain tissue, the high-power executive branch, so to speak, of the mental government. Obviously it cannot reenliven dead tissue, but it can reanimate tissue that is simply loafing but still intact. The primitive organization fights viciously to maintain control of the brain and thus, in its pre-verbal ways, ignore higher brain structures. If we can block the hyper-dopaminergic flow and restore gating, we can once again enliven those higher structures that have yet to deteriorate. This is why some schizophrenics are Yale graduates and Nobel Prize winners. So why not use clozapine as a first-choice medication to try to preserve as much brain tissue as we can? The stated risks and the swamp of side effects and required blood tests prevent us from doing that, with the FDA insisting we try at least two other medications first. The goal in schizophrenic treatment is to maintain brain tissue, and by juggling the dopaminergic tsunami with our chemical potions, we can do just that but as that tissue deteriorates, response to medication wanes. (It should be noted that schizophrenics pay for the accomplishment of language skills with the ultimate deterioration of their forebrain. This area has learned to accept conceptual input with the help of dopaminergic suppression as the normal modus operandi. Language skills are promoted by parental interaction, schooling, reading, and writing, all of which gear us up for a life of brain expanding conceptualization. For the schizophrenic who masters language with a bard's skill, this falls apart in late adolescence. The primitive organization reclaims the cranial territory, and things go sour. However, dedicated efforts to get the schizophrenic to use the executive brain loci do yield results and are protective.)

Clozapine became the treatment of the moment. In their uniqueness, clozapine patients are enrolled in a special program so no one is given their weekly dole of drug without having their blood drawn. The perilous white count plunge can be picked up quickly enough to stop the drug and allow the count to renormalize. But that's it. We rarely reintroduce clozapine once the dreaded blood count plummet is encountered. Deaths are very rare but side effects are not.

Clozapine specialized in being oddball as it didn't block just dopamine receptors; it blocked another receptor called *serotonin 2a*. In addition it has the property of clinging and de-clinging to allow a bit of dopamine to dribble through. If the goal is the blockade of dopamine, clozapine seems to do it better. Unlike other antipsychotics, clozapine tends to leave the nigrostriatal tract alone, reducing the incidence of tardive dyskinesia. Another thing in its favor is not causing neuroleptic malignant syndrome much either. So it seems to be an advance in dopamine restoration, restoring the previous neurochemical balance better than any other. And it just so happens that psychedelics trigger that very same serotonin 2a receptor, whereas clozapine and the other new, atypical neuroleptic potions block it more than dopamine receptors themselves.

Of course, as pharmaceutical companies did with Thorazine, they started churning out clones of clozapine, leaving out the part that causes white cells to plummet. None of the new miracle cures like Olanzapine, Risperidone, Quetiapine, Ziprasidone, Aripiprazole, and many others caused this nasty, sudden, white-blood-cell-count dive and therefore, so there was no need to register in a program or have a weekly blood test. This is not to say that they didn't have grizzly side effects. The early clones of clozapine all caused ballooning weight gain and sedation, and some, like Risperidone, cause an increase in a brain chemical called *prolactin*.

This hormone normally induces breast swelling and milk production (lactation) in pregnant women and can do so in male and female patients —an unwelcome discovery along with the occasional impotence that tags along with it. The even later clones of clozapine have avoided the weight gain fiasco but have other nagging nasty habits and have names like Ziprasidone (which can have serious heart rhythm side effects), Aripiprazole, Asenapine, Lurasidone, Iloperidone, Cariprazine, and so on in the chemical congener dance. So the psychiatrist has a vast array of cards to choose from all vying to recoup their staggering production costs for their patent, and all of them, like the characters in a Cagney movie, imperfect. The goal is to find the one that fits the patient like a comfy suit, affably informal, reducing symptoms maximally with as few side effects as we can get away with. Yet psychiatrists are in the position of needing more effective powerhouses even than clozapine. They don't do enough reliably and pleasantly without goblinesque side effects to be the noteworthy salves we need and our patients deserve.

A fictitious case example might demonstrate the uniqueness of our side-effect-ridden, super-boy-potion clozapine, the gold standard of antipsychotics. Cole is a paranoid schizophrenic. While he takes his current medication (Olanzapine) regularly, he has yet to understand why they insist on this nuisance and has ballooned eighty pounds while registering towering blood sugar and skyscraping cholesterols. He's kind of getting by, hanging on to a crushingly boring, low-challenge job and floating along with the patients he shares his group home with, but he has started to demonstrate troubling behavior. Some feistiness crops up around his peers, and he's told the staff that trains were sent near the house just to rankle him. He accused two innocent housemates of theft and threatened to punch one of them. He has stopped eating, accusing the the kitchen staff of wanting to poison him. His demeanor is suddenly funky and withdrawn. He lurks around in a dirty long blue coat even in blistering heat, and he's been caught drinking his own pee. Cole shrugs it off and withdraws even more, settling into the cocoon of his own head space. Later he brags about curing himself of cancer.

His psychiatrist, Dr. Granger, recommends a stay at the local hospital, given the threatening nature and deteriorative presentation of his patient, to which Cole surprisingly agrees. Dr. Granger recommends the inpatient doc make a switch to clozapine and forwards a condensed summary of recent events. The clozapine was ratcheted up while the Olanzapine was tapered.

Two months later, Cole shows up at Dr. Granger's office looking rather spiffy and alive. His face breathes some affect, and he's even joking around a bit with an air of nonchalance. He couldn't care less about trains and he's not wearing his blue overcoat. Occasionally he makes a puzzling statement but doesn't feel that anyone at the home has it in for him. Even his family thinks he's back to his old, self-absorbed self. Complaining of drooling and constipation, he accepts the side-effect medications Dr. Granger offers him.

These are some of the typical gains we see with Clozaril. They may seem rather piddling but they can be huge, and some patients will have much more dramatic advances. Patients with longtime schizophrenia, generally labeled "residual type," will improve in overarching demeanor and soar to higher heights of functioning than they had before. If clozapine fails to deliver them to mental peaks, then ECT might be the next recommendation. Many of what we call *high-functioning schizophrenics* just so happen to be on clozapine. It just works on a different level, perhaps even focusing on mesolimbic and mesocortical dopamine tracts more than the other antipsychotics. Clozapine seems to have a knack for leaving alone the nigrostriatal tract, which could be due to its strong attachment to the serotonin 2a receptor and light touch when blocking the dopamine 2 receptor. Of note, more patients seem to develop obsessive compulsive symptoms on clozapine. This at first glance might seem like a bad thing, but it could be precisely clozapine's success at restoring the premorbid neurochemical balance that encourages patients to use obsessive compulsive symptoms in an attempt to further suppress dopamine by promoting acetylcholine.

So now you understand these designer medications (antipsychotics or neuroleptics) almost as well as the desperate psychiatrist. The doctor soothes a mysterious illness he really doesn't grasp with medications he has only a fleeting rationale for in patients who typically are in full-court denial that they have any condition let alone a heartfelt desire to take any side-effect-laden potions for an illness that they assume is imagined by their controlling family and a money-hungry MD. It is not surprising that schizophrenics many times just stop their medication and subsequently free-fall into full-blown psychotic decompensation, exposing themselves to searing and irreversible brain atrophy. The doctor becomes talented at assessing the results of his efforts, the benefits, and the significant side effects. Yet one could truly say he is flailing in the dark, no doubt good-naturedly hoping for success in a blind man's endeavor. No one in medical school or beyond has ever given him an explanation of who these suffering people truly are, from whence their affliction emanates, and what the medications actually rectify . . . until now. Its much more comforting to know the underlying cause of an illness when treating it than to deal with the unknown, which also leads to inflated stigma.

* * *

What does the psychiatrist do when evaluating someone who might possibly have schizophrenia without the guiding beacon of lab tests, X-rays, or physical abnormalities for proof? Well, age of onset is certainly one clue, late teens to early twenties being the most usual span. But there are exceptions to every rule, and I treated a patient a while back who was in her early sixties when her illness began. She had no mental health intervention previously but began proclaiming that the devil was inside of her, and she was inestimably adamant, wailing and moaning, that Mephistopheles had taken up residence somewhere in her body. She went to a church and demanded the priest remove the grizzly intruder, pleading for exorcism, and her family became afraid of her and called 911. In the hospital she was fawningly compliant with treatment but there was a strong feeling that she was actively camouflaging her real thoughts. A thorough neurological work up was negative, dismissing any chance of tumor, dementia, infectious disease, or autoimmune illness and no evidence of anything but schizophrenia. Gradually she worked her way toward amelioration on a medication called Latuda and got discharged from the hospital. Within a week she had to be brought back because she again was thinking that the devil was hiding in her closet and was even more beside herself when told this was nonsense. She started drinking heavily and when she got back to the hospital admitted she hadn't been telling us everything including that she was spitting out some of her pills, or "cheeking." (While we can monitor for this, some patients get supremely talented at it.)

Schreber was in his forties when he had his first psychosis yet he was clearly schizophrenic. Some would dispute this but he had obvious paranoid delusions, hallucinations of bees attacking him among other things, bizarre extremes of behavior including catatonia with statuesque rigidity, physical brutality, and more. He met every DSM-V criteria for schizophrenia. His illness started when he woke up from a momentous dream one morning with the alien notion that it "really must be rather pleasant to be a woman succumbing to intercourse" (Schreber 1988). This was the key to Schreber's handling of the takeover by the insidious primitive organization . . . He viewed it as a sexual submission and indeed sexualized the whole landscape of his nightmarish insanity. Let's review the Schreber case and see what, if anything, we can glean from it.

Schreber was a German judge who in the early 1900s endured a late-life psychosis that was clearly schizophrenic. A fastidious, intelligent man, he left an intimate account of his madness in *Memoirs of My Nervous Illness* along with his own interpretation of events. Many people have given their opinion of what Schreber went through, and much has been written discussing it.

Before the onset of their first major breakdown, some schizophrenics get an inkling that their minds are on the verge of an unstoppable encroachment. This frightening and unnerving conviction causes a variety of counterreactions, what one might label on-the-way-to-psychosis phenomena, that are likely attempts to forestall the inevitable. (Indeed, one could advance the postulate that all mental illnesses are a concerted effort to sidestep the schizophrenic extreme, the seismic entropic victory. In this playbook, the other nonschizophrenic diagnoses have succeeded in damning up the dopaminergic flood before it reaches its ultimate saturation point.) Examples of this in more modern literature are all over the place. See, for example, Saks (2007), which describes a harrowing major depressive episode before the initiation of full-blown, florid psychosis.

Schreber suffered an earlier infirmity that was characterized as hypochondria and that eventually resolved (Schreber 1988). For the next eight years Schreber lived a competent, pro forma life. Then, he writes, "I dreamt several times that my former illness had returned." He was ticklishly relieved that these were just dreams, but not long after, he had the sensation that it "really must be rather pleasant to be a woman succumbing to intercourse." He was so masculinely indignant about this notion that he concluded "external influences were at work to implant this idea into me." In short, he instantly leapt to the supposition that this couldn't possibly have arisen from his own mind.

What he was describing was part of the premonition that his illness was about to resurrect full force; however, this new second affliction would now be nothing less than schizophrenia. The threat demanded a strategy and especially in someone as studiously bright and obsessively concerned about their health as was Schreber. Now his mind's unconscious tactic was to submit to this force (the primitive organization) as a woman submits to intercourse, odd as this sounds, thus deriving some source of pleasure from this feminine stance and perhaps weakening it. The feminine impulse was no doubt a gnawing facet of his unconscious playground long before the primitive organization's coup. It might have been present in his father as well, a man described as a major proponent of male exercise, a macho man's man, which was arguably a sublimation of feminine impulses. At the same time Schreber's diffident stance offered a desperate feeling of partial control over the oncoming mental tsunami, and he needed "a wall in protection of God's realms against an advancing blood tide" (Schreber 1988). The oncoming illness was clearly viewed as an assault with multiple metaphoric references to its onset including "soul murder" and "unmanning" with the belief that "the whole of mankind had perished," referring no doubt to himself. He even read his own obituary. All of these symptoms marked the onset of an eight-year-long traipse through a psychotic wonderland.

This tantalizing, intrapsychic ballet allowed Schreber to gratify something that had been sticking under his unconscious craw for years: forbidden feminine wishes. In the light of an oncoming monsoon, he was able to dance merrily in the rain, soaking in the torrent of his forbidden female desires. This mastodon-like psychotic process was the godsend his unconscious female yearnings longed for, granting tacit permission, thus greasing the wheels of this oncoming tide. And of course the deflation of this mental bondage released a surge of repressive energy, granting entropy's desire.

Implicit in the unstoppable uprising of the primitive organization is a full-scale cognitive regression to a point in time before language, allowing for magical and childlike thinking. Niederland (1974) refers to the "ego's outspoken disorganization in psychotic illness. In such states, fantasies, experienced as real become delusions." For example, Schreber was now convinced that his body was to be handed over in the manner of a female harlot (Schreber 1988). Early on he attempted suicide, begging for poison while at the same time dreading being killed. He both enjoyed his feminization and denounced it, finding the means to sexualize the experience while maintaining a desperate masculine protest against it. According to Freud, Schreber's doctor, Flechsig, was the object of Schreber's sexual desire, transferred from his father and brother. Schreber experienced Flechsig's soul in his body, a bulky bundle with a "foul taste and smell" (a symbolic reference to semen). Schreber delighted in the "sensation of voluptuousness" so much so that he "gained such a strong power of attraction," giving his body a "feminine stamp." In short, one could say that Schreber took the occasion of the onset of his schizophrenia to fulfill his probably long-standing, unconscious wish to be female.

Schreber viewed this as a choice between two evils, and the justification for choosing "feminization" was presumably to preserve his reason. "I would like to meet the man who, faced with the choice of either becoming a demented human being in male habitus or a spirited woman, would not prefer the latter" (Schreber 1988). Thus the clever barrister, confronted with a momentous inner crossroad, went with femininity and reason. It became "imperative even during meals to furnish the distant God with proof that my mental powers were intact." If he was to lose his genitals and the dressings of his masculine arrogance, he was definitely not going to forfeit his sizable mental intensity. Aware of the famous psychiatrist Kraepelin's "dementia praecox" (meaning early dementia, a pejorative term Kraepelin coined for schizophrenia), Schreber eschewed a dreaded dementia at all costs. "The whole policy God was pursuing against me aimed at destroying my reason had failed." In short, he chose his mind over his genitals.

The primitive organization's childlike rules of logic also allowed him to attribute his experiences to religious miracles, eschewing the loutish schizophrenic label. "I have no doubt that my early ideas were not simply "delusions" and "hallucinations" (Schreber 1988). In fact his entire literate memoir is a textbook, so to speak, of God's properties and intimate mechanisms. Despite his denial, Schreber clearly demonstrated the classic symptoms of schizophrenia including visual and auditory hallucinations, tactile hallucinations, delusions of a religious and sexual nature, catatonic staring, and bizarre behavior, and his condition met every diagnostic criteria for schizophrenia in DSM-V although its onset was atypically late. Schreber saw himself as the rotational fulcrum of the universe, a viewpoint consistent with childlike, self-centered thinking: "My person has become the center of divine miracles." "Everything that happens is in reference to me," he writes, including the weather, planetary alignment, and attacks of wasps.

At the same time that he indulged himself in the forbidden sexual reward of his new state of "voluptuousness," he also viewed himself as a divine teacher, predicting his own fame not as a patient of Freud's but as an interpreter of God. Tortured by positive symptoms, his mind space was inundated by derogatory, low-pitched voices from hell, repetitive phrases, and the need to constantly conjecture. (This is the result of the dopaminergic resurgence that floods the brain space and nullifies the gating function, thus allowing the primitive organization to surge. This mind-brain invasion is a common feature of psychosis of all types.) These hatter-esque phenomena were attributed to nerves and rays. He projected his own diminishment by the primitive organization onto those around him, whom he described as "fleeting improvised men" and "fossilized men" (Schreber 1988). He spoke at length of God's prurient attraction to him and experienced "mental torture" from which he "suffered severely for years"—no doubt a product of the combat between the primitive organization and what was left of his adult mind. At one point he hallucinated that he had no abdomen at all, his food pouring directly into his thighs.

This degree of invasive insanity is entirely typical of schizophrenic thought patterns. Schreber was so violent with other patients that he lived in a padded cell for over two years, at times requiring forced feeding by attendants. His masochistic maladies multiplied to include lumbago, toothache, and headaches. Here we see intense positive symptoms indicative of the conflict between Schreber's lumbering mentality and the primitive organization. These symptoms would have largely cleared up if medication had been available at that time but it would be another half century before Thorazine made its momentous appearance. Our ego is the command center that has dragged us away from our primitive, 1.0 Neanderthal past and into modernity with its civilization, religious values, conceptual skills, and 2.0 rules of adult thinking. In order for the primitive organization to succeed in returning us to the pre-verbal, pre-civilization state, it must vanquish the ego and deflate gating. Eventually the ego succumbs to this overwhelming, relentless barrage of denigration, but first sparks must fly in the form of positive symptoms. Once our friend the ego collapses, old entropy dances its final fiendish jig.

Eventually, however, the voices decelerated (Schreber 1988). "The main danger which seemed to threaten me during the early years of my illness is removed." His condition underwent a gradual denouement directly proportional to the extent to which he abandoned his masculine protest against his wish to be a woman. "I must submit to the necessity of the order of the world which forces me to accept these ideas to avoid pains." The primitive organization embraced the battle cry of dissolution of all resistance to feminization, which coincided with a relentless derogation of Schreber's floundering ego. With this accomplished, the intensity of positive symptoms abated as Schreber observed a "slowing down of voices" and that "miracles are increasingly harmless" as "God's hostile opposition to me continues to lose in virulence."

Schreber languished in the imagined benefits of his torturous odyssey including nothing less than immortality: "Ordinary illnesses even external violence cannot cause my death" (Schreber 1988). He anticipated "great fame," which oddly enough he achieved as the basis for Freud's incisive case study of paranoia. His inflated grandiosity allowed him other imaginary blessings: "A very special palm of victory shall be mine." These blessings permitted him to justify in his own mind a hesitant acquiescence to the feminization he initially disavowed. The primitive organization, having run its course, evolved into a more conciliatory condition, melting away the positive symptoms. This typifies the natural trajectory of the battle-weary schizophrenic process, which often leaves its victims grimly saddled with a plethora of negative symptoms (illogicality, lack of pleasure or motivation) coinciding with the attenuation of the more dramatic, positive manifestations like voices, bizarre behavior, and florid derangement of speech. Thus the mind state of the residual schizophrenic can be said to be entropy's cherished endpoint short of a suicidal return to inorganicity itself. Negative symptoms broadcast total ego disenfranchisement coupled with blunt neuronal atrophy and the embodiment of full dopaminergic non-suppression. This cerebral collapse could not have taken place before language's dopamine-suppressive potency, meaning that our Mr. Neanderthal could never have experienced the mind-wrenching entropic state of the schizophrenic. The battering ram of dopamine suppression teed up the groundwork for an earthquake of de-suppression greater than anything that could have been anticipated before language.

Schreber's approach to the oncoming primitive organization may be a typical feature of paranoid schizophrenics. They are known to have a delicately better trajectory than other schizophrenics perhaps for the reason illustrated by our friend Schreber: the fulfillment of a prohibited sexual gratification in exchange for less cognitive damage.

The primitive organization is by definition cognitively regressed. It derives from the pre-verbal state our Mr. Neanderthal and his brethren endured for some six million years before language when the landscape of consciousness was experiential and we perceived what was dished up by our perceptual organs and our raging moods, our fleeting memories. We had little control over the weighty pumpkin atop our necks, and helpful as it was, our brains functioned on autopilot without consulting us. Over the past fifty thousand years or so, the ego has jettisoned humankind forward, with the blessing of language, in cognitive development and gained control of the theater of consciousness with words as its tool. This incredible feat was no doubt driven by natural selection and its clear endorsement of cognitive supremacy. If you take a five-year-old and a twenty-year-old and let them loose in the jungle, even if they are physically equal, the latter is more likely to survive given his mentally advantageous complexity. In the same way, the ego pulled the locus of human brain functioning away from the primitive organization like a mule bucking on a rope while borrowing some of its energy. This process is re-created during each individual's maturation with the fluttery movement away from one's internal preoccupations toward the sharply delineated outlines of external reality.

When the primitive organization comes roaring back in schizophrenics, those gains scatter. One could say that the flagship of mental illness is a loss of the ego's hold on consciousness as the universe of internal idiosyncrasies dilutes the foundation since the gating function of dopamine suppression is lost. Reality testing is compromised big time, and the schizophrenic returns to the inner experiential world of the child subject to all sorts of psychotic phenomena in his or her regressed condition. Freud, despite his brilliant discussion of the paranoid mentality, sidestepped the glaring canvass of Schreber's schizophrenia. Paranoia, he argued, can be explained by the projection outward of homosexual love onto another and then reversing it to its opposite, hate. Yet there is a difference in thinking between, for example, "Maybe my doctor is out to get me" or "Maybe my boss is trying to get me to make a mistake" and "My doctor is making a clear effort to end my life." The latter flaunts the marrow of reality testing in a wholesale distortion indicative of primitive thinking. In fact, Freud never took it as his task to explain schizophrenia at all yet in the world of prehistoric schizophrenic thinking anything is possible.

Schreber's bargain, his mind or his genitals, could only happen in a brain barely tethered to any reality. Our ancient mindset knocked on his door, a door weakened by his unconscious wish to be a woman, a distant yearning now endorsed by the encroaching primitive processes of thought. Not even one as massive in brain power as Freud took into account the persuasiveness of his invasion, which was entropy's desire, the ancient constellation of a Neanderthal's brain. Yes, paranoia has roots in various impulses but all are fueled by the gas of the primitive organization in the schizophrenic and the loss of repression when it takes over, in response to the de-suppression of dopamine. If Schreber's father complex made him hate his doctor, then primitive thinking underlined the project. The newly resurgent caveman brain exploited all Schreber's foibles and idiosyncrasies, dragging his mind into a delicate torture and then ultimately easing with his acceptance that he was now, indeed a woman—a small price for him to pay for what was left of his complex sanity.

Here we see the necessity of the viewpoint that schizophrenia is a replacement of brain function by a cognitively primitive state carried over from our ancient prehistory. It allows for a complete overview of the Schreber panorama and explains why some paranoid schizophrenics do better than others. This resurgent, upstart mind and brain has its origin in the timeworn evolution of our forebears using dopamine de-suppression and delusional thinking to achieve its goals, all stoked by entropy's enduring seduction in the context of an all-too-recent evolutionary metamorphosis.

We can see from the Schreber case that the symptoms in schizophrenia run the gamut of creative insanity. Each schizophrenia is different. Each person puts their own stamp on the dance and in turn the interactive dance on them. Those symptoms generally fall under what we call *psychosis*—that is, noticeable craziness, a term that is a general indicator for thinking and behavior that has lost touch with reality, is not driven by adult modern logical motivation, and is noticeably off-kilter compared to the behavior of most adults. If I run through the streets flapping my arms like a bird, I'll probably get the attention of others and ultimately the police who will haul me to the emergency room. There are quiet, genteel psychoses as well, people who are delusional or hallucinating and do not draw much attention to themselves. People can be unhinged mentally without any legal intervention. It's only when they become a full-throttle public nuisance or a threat to self or others that the law intervenes. A patient can come to an emergency room and claim that aliens have landed in his attic and are building a time machine so that he can be transported back to Cleopatra's boudoir and yet be released if he doesn't want to stay in the hospital and is not a danger to himself or others. Nor can we force them to take medication since they are not in the category of threat. (We might still offer it to them though with a strong dollop of encouragement.) On the other hand, if these beliefs prevent him from fulfilling his basic needs for food, shelter, or medical care or place him or others in an obvious peril, then it's a different game. One would then make a case for hospitalization and either implore them to take medication or petition the county to set up a civil court hearing to force the issue. Schreber ultimately went to court and won his freedom but only after years of painful battle with the primitive organization that turned his mind into a house of horrors.

If you are attempting to explain schizophrenia then you must be able to explain that rather vague term "psychosis." Perhaps this sounds easy. After all, if you take LSD, you become psychotic, or if you use methamphetamine you become crazy and often in ways that are indistinguishable from paranoid schizophrenia. I use these examples to point out that there are what one might term "serotonergic psychoses" related to hallucinogens like LSD that have a different quality from the garden variety dopaminergic psychoses due to stimulant drugs like methamphetamine. The former are more introspective and tend to produce more visual hallucinations yet unfold from the same jungle as their stimulant variety cousins—namely, de-suppression of our central player dopamine. Stimulant psychoses rustle up more obviously paranoid symptoms in which people believe their computers have been hacked by pornographers and CIA drones are monitoring their every move or that voices are blaring from walls that harbor clandestine bugging devices. So it turns out that there is a psychosis related to stimulation of 5HT2A serotonin receptors, which indirectly jettison a dopaminergic torrent or by the prodding of D2 dopamine receptors by direct release of the stuff. (The hallucinogens ayahuasca and 5-MeO-DMT found in the Sonoran desert toad stimulate 5HT2A receptors, which no doubt leads to a crescendo of dopamine de-suppression. LSD is itself a strong stimulator of dopamine receptors, being a chemical look alike of serotonin. With these so-called psychedelics, the ego seems to vanish and one regresses to a state of internally preoccupied infancy.)

So stimulating 5HT2A receptors indirectly undoes everything that evolution with the help of language has done over the past fifty thousand years. It also proves that a prehistoric, brute, primitive organization resides in us all and it just takes the right molecule, born of plant or man, to awaken the grouchy giant. Great, but unfortunately this doesn't explain much about the psychosis of schizophrenia. No surprise that the spiffy new atypical antipsychotics aren't just the recycled old stuff but bow to two masters, a stodgy old dopamine receptor and a serotonin 2A receptor to boot. One might dub them un-psychedelics, more potent and with a variety of side effects in a profile different from the old, gruff, standard antipsychotics. The psychotic dance remains in us all. Rather than overthrowing the 1.0 basic mindset of our Neanderthal progenitors, we grew a new layer of mental acumen over it. A neurochemical process has guided us to the promised land of a heightened mentality, delicately reversible, slyly silent until sleep awakens it or a bitter cup of ayahuasca brew waves it forward using the process of dopamine de-suppression touched off by serotonin 2A's roundabout genius.

We need to divide psychosis into its parts. The ego's roles include functions that coincide with the suppression of dopamine: fostering gating, prioritizing incoming stimuli, testing reality, separating self from others, creating an awareness of death and the continuity of time, and delivering concepts to those high-power, executive brain centers, the dorsolateral prefrontal cortex. When the primitive organization reasserts itself, the ego takes a hit. This results in a cognitive backslide to thinking that is childlike and plays by different rules, which explains a lot of the expressive oddities of schizophrenics. They're thinking with simplified, more-diffuse, childlike habits that make no sense to us and sound crazy. The observing mind of the schizophrenic, whatever is left of it, acknowledges some mysterious paradigm shift in cognitive functioning without understanding it. Cognitive regression is therefore christened a major facet of the psychotic panorama. The ego's function of reality testing is massively impaired, which opens the schizophrenic up to delusions and oversimplified misconceptions. As we saw in the Schreber case, delusions are largely determined by the psychodynamics, the unique individual issues of the patient. In Schreber's case, he suffered a father complex, according to Freud, and had sexual yearnings contorted into persecutory thinking regarding authority figures. This, as Freud asserted, explained some of Schreber's symptoms. The ego, also in charge of differentiating ourselves from others, crumbles, and this leaves us in a free-floating sense of universality. The nascent flood of de-suppressed dopamine dissolves the gating function and allows intrusions into the vaudevillian theater of conceptualization leading to a perceptual cacophony. Add to this the creation of hallucinations by the sensorily deprived forebrain, and the situation becomes very chaotic. If we lose the ability to organize incoming stimuli, the putative gating function, then it feels as if everything coming into our brain is of equal importance and there's no way to grasp it all or to determine which stimuli are most important to survival.

This is a good time to discuss prepulse inhibition. The schizophrenic's attention span is diffuse, helter-skelter, and lacking focus in comparison to modern man's and is childlike in that regard. The phenomenon of prepulse inhibition (performed in a lab under controlled conditions) involves presenting a loud stimulus to the subject (like a gunshot) through headphones. Of course most people will be startled when they hear this. If, on the other hand, you present the subject with a dull, thud-like noise, or prepulse, just a fraction of a second before the gunshot, the subject's reaction to the gun blast will be diluted and inhibited. Enough of his dense attentional locus is grabbed by the dull thud to reduce or gate out his startled response to the piercing gunshot.

Now consider that we are testing in this situation a person's level of cognitive organization. Well, it shouldn't surprise you that schizophrenics demonstrate an abject failure of this prepulse inhibition. When presented with the dull thud sound a fraction of a second before the sharp, loud gunshot *blam*, their reaction is largely unchanged. Why? Visualize your attention span as a circle with dots in it. For modern humans, that circle has many tightly packed dots. When a dull thud is presented to a modern sapiens, a large number of those dots is captured by that noise leaving fewer to respond to the loud gunshot. If, on the other hand, your primitive attention span 1.0 consists of far fewer concentrated dots, fewer dots are captured by the dull noise, leaving much of it intact to respond to the gunshot noise, and since their gating function is impaired, they don't gate out the gunshot while attending to the prepulse. In short, schizophrenics fail to demonstrate prepulse inhibition because their focus is more diluted and vague, their gating impaired. Surprise surprise, so do children and people taking LSD, groups that also have a more-diffuse, regressed, less-focused, and less-gated field of attention.

And consider this. For humans to gate incoming stimuli, there must be a delicately refined ranking protocol of evaluation that takes place within milliseconds of experiencing each stimulus. The mind will need to rate these sensory signals, let's say on a scale of 1 to 10, in importance in order to decipher which ones to grant entry to the royal palace. A 10 demands our full attention and a 1, well, forget about it. When the normal test subject hears the dull thud, this micro-evaluative process is rolling forward and blocking other incoming stimuli just as the gunshot blares through the headphones. In short, we're busy with prioritizing and so we gate out the louder noise. Thus a failure of this prepulse inhibition represents the mind's scattered, diffuse, 1.0 locus of attention and lack of gating of incoming stimuli due to a torrent of unsuppressed dopamine. This corresponds admirably with our theory about schizophrenics' diluted level of thought organization, lack of gating, and ego dissolution. And by the way, when schizophrenics are medicated with antipsychotics, of course, they normalize in this regard.

So what is Charles Bonnet syndrome, and why do we need to talk about this, and why now? Old age, a harbinger of sensory decline, foments a condition of disquiet in some called *Charles Bonnet syndrome*. We lose our vision and hearing for a multitude of reasons: age-related macular degeneration, for example, or simple auditory fade. When this happens, the visual cortex and those other parts of the brain that are used to input from the eyes and ears start to self-stimulate because disuse leads to the dreaded atrophy of tissue, which is a one-way street as dead nerve tissue does not rebloom. They do so by creating their own input, called *hallucinations*. Why, you might ask, does this happen? Well, if a muscle stops getting the usual stimulation such as in paralysis, it begins to twitch so as not to deteriorate, and we counter this dire state of affairs with externally applied movements. Similarly, brain tissue, without stimuli, will atrophy and so it tries to address this abject lack of input by creating its own show as it were. Hence elderly people may see things as they lose eyesight or hear things as they lose hearing. And these are not people afflicted with any mental illness whatsoever.

So what does all this have to do with schizophrenia? If, as I've suggested, the primitive organization storms the brain and the higher centers are short-circuited, then we observe self-stimulatory phenomenon, echoey voices and phantasmic visions, because the executive centers are deprived of their usual meal and in fact interpret this as a state of sleep. This explains the expected haunting intensity of hallucinations in schizophrenics. You recall, of course, how the spiffy, modern, 2.0 brain is fed conceptual input streams by the ego, which has become, in essence, a sixth sensory organ, peppering in thought missiles. We learn, as we mature, to supply the prefrontal lobes with waves of incoming snacks so that the executive centers of the brain can munch on them in a contemplative way. This actually becomes the modus operandi for the brain's invigorated functioning and is accompanied by the secretion of the brain fertilizers BDNF and VEGF. The primitive organization, being pre-verbal, does not do that, and the frontal lobe with its dorsolateral prefrontal cortex is short-circuited, as modeled in sleep. In an attempt to shoo away the dreaded atrophy (which is well known to occur in schizophrenics over time, worsening with each relapse), these areas begin to self-stimulate, producing their own Broadway production. They create their own input in the form of hallucinations. When else does this happen? It's called *sleep*, and when we sleep, a large share of sensory input is ratcheted down. Eyes closed, little noise, comfortable temperature, muscles paralyzed since the neuromuscular input disconnects (since muscular tension requires dopaminergic suppression), there is vastly reduced sensation, and we de-suppress dopamine flowing like an inner tube down the river to a state of maximum entropy, the brain's restorative salve. So how does the brain respond? It hallucinates, in the form of dreams, a self-stimulatory phenomenon during the fog of sleep. The analogy between schizophrenia and a waking dream state entertains some credibility.

But hallucinations, you will say, occur under other circumstances. There's a whole class of drugs called hallucinogens or psychedelics (now you know why). I've mentioned some of them: LSD, DMT, 5-MeO-DMT (the Sonoran desert toad excretion, one of the most potent psychedelics known to man), shopworn cannabis, which unfortunately is being legalized recreationally all over the universe despite its proven harm to nerve cells especially in teens (people using cannabis often don't hallucinate, possibly because of CBD, another marijuana ingredient, which has some marginal antipsychotic properties in high doses. So when people smoke or ingest marijuana they are imbibing a hallucinogen with a mildly potent antipsychotic), mescaline, psilocybin, and the list goes on. They are known to cause hallucinations, and it turns out that even these serotonin receptor 2A provokers, the self-same receptor blocked by the newer antipsychotics, evince some intense dopaminergic de-suppressing properties resulting in an instantaneous ego shredding as the psychonaut regresses back to the primitive organization abandoned by dopamine suppression long ago (De Gregorio et al. 2016) as are stimulants, drugs that directly surge dopamine, like methamphetamine and cocaine. According to my theory, all of these have in common a short-circuiting of the higher centers of the brain and hence, a self-stimulatory, hallucinatory manufacture, the waking dream, in an attempt to prevent atrophy.

All of this confirms my theory that anything suggestive of craziness that we label psychosis is related to dopamine excess. And furthermore, could it be then that the transformation from the physical and experiential grunts called cavepeople to the complex, latte-guzzling modern with his or her conceptual consciousness 2.0 requires a gradual suppression of dopamine? Right on both counts. So let's imagine that as we mature we use secondary process thinking (with the help of child rearing, years of schooling, reading, etc.) by suppressing the dopaminergic tone of our early years, the very same dopaminergic tone present for millions of years in primitive man and the same dopaminergic release revisited nightly in slumber. The primitive organization throws that tone into full throttle reversal perhaps just as it itself ripens to full biological development (and the effort of gating enlarges to the breaking point), hence the onset of schizophrenia in one's late teens, early twenties. For drug users, they usually return to the proper dopaminergic tone within hours as the psychedelic brew leaves their Woodstockian system (but not always as methamphetamine and other drugs can at times cause a permanent, poorly treatable psychosis). Schizophrenics don't return to that dopaminergic suppression and require antipsychotics to do so. Further proof that dopamine is suppressed once we move out of adolescence lies in stuttering and Tourette's syndrome, both of which tend to occur in childhood and often gradually fade as the individual reaches his or her late teens, early twenties. And is it any wonder that both respond to dopamine-blocking medications?

This habit of binding dopamine, our star neurotransmitter, received its first modeling in the motor system, the nigrostriatal tract, where it had to be harnessed to fall into line with another erstwhile neurotransmitter called *acetylcholine*. Failure to so do would result in the ungainly movements of, say, Parkinson's disease, where there is a decided lack of dopamine, or alternatively Tourette's, where the dopaminergic tone is overbearing, forcing its partner in crime acetylcholine to punch its way through with an involuntary jerking movement we label a "tic." So the prototype of this dopamine lockdown was the motor area, where as kids we learn to coordinate our fastball or integrate our crocheting with repetitive movements. The intricacy of speech is a most decidedly suppressive endeavor involving multiple muscles and organ systems like tongue, lips, teeth, diaphragm, and vocal cords coalescing in a grand orchestration to warble out the twenty-six letters in our alphabet. (Neanderthals couldn't produce nearly this many.) And speech is intimately involved with thought; thus our dopamine-suppressive principle transitioned from the motor area to the conceptual realms, the mesocortical and mesolimbic tracts. Bingo! We were off and running unlike any other species of Homo, the poor Neanderthals, the sad Naledi, the mopey Denisovans. Only in their dreams could Neanderthals reap the ravishing rewards of language as we did.

Freud theorized that it falls upon the ego to control voluntary musculature. With language, our ego received its grandest promotion, becoming puppet master of speech and thinking and syntax and the wording of everything. The same ego appointed to ride roughshod over our movements oversees the purview of speech and ultimately thinking, relying on the feat of dopamine suppression with all its benefits.

So let's see if we can list the mechanics of what we call psychosis. We want to separate generic psychosis into its component parts.

1. Cognitive regression. Schizophrenics and other psychotic people don't use the same rules of logic as the rest of us. Instead, they rely on primitive, childlike rules (Werner 1948; Werner and Kaplan 1963).
2. A mammoth surge in dopaminergic tone either via drug use or the return of our Neanderthal primitive organization 1.0. (This also happens in REM sleep, which brings on hallucinations in the form of dreams.)
3. Short-circuiting of the higher centers of executive functioning, the Sherlockian deductive reasoning locus in the dorsolateral prefrontal cortex, leading to self-stimulation with hallucinations à la Charles Bonnet.
4. Potential atrophy of those same higher centers of the mind from abject disuse over time, worsening with each psychotic relapse, and the lack of those enriching brain stimulators BDNF and VEGF. This is why we need to diagnose and jump on the schizophrenia treatment bandwagon in a timely, aggressive manner.
5. A personal psychodynamic that can guide the direction of symptoms, as in Schreber's florid paranoia.
6. Destruction of vital properties like gating with prioritization of incoming stimuli, reality testing, and ego fortitude. Hallucinogen users have described this feeling of ego shredding with distinct clarity.

Chapter 7

Evolution versus Entropy, the Theory in Its Totality, a Look into the Future

In his book *Beyond the Pleasure Principle*, Freud reiterated his broad-based theory that the mind avidly seeks pleasure and avoids unpleasure. While this is not such an earth-shattering declaration, he defined precisely what he meant by this: "Unpleasure corresponds to an *increase* in the quantity of excitation and pleasure a *diminution*" (Freud 1961). In so stating he was making reference to the variegated symptoms of his patients and the addictive yen on the part of the mind to minimize unpleasant excitations. The mind did so by burying unwanted impulses, such as those forbidden by societal taboos, but this depth of burial also captured energy and effort. These impulses, for example those nasty Oedipal ones that lead to human tragedies like patricide or fratricide, would be pushed far from their word labels, plummeting into the unconscious depth as would the primitive organization en masse. So now in its effort to limit excitation, in the spirit of the pleasure principle, the mind actually increased its energy with these assignments born of civilization's need for order. This wholesale burial demanded the intensity of Thor and in so doing, an exertion of excitation. Furthermore, as the repressed impulses tended to seek some kind of gratification, the mind occasionally gave in and created cleverly crafted compromise satisfactions while still keeping the banished impulse out of awareness—in short a Herculean effort. This led to a host of symptoms that drove Freud's patients to clamor for his singular care. If Freud could bring the buried impulses into awareness, he could set free the tension overall and the need on the part of the mind to manufacture these inventive symptoms. Doing so led him to something called *transference*. I hypothesize that lifting repressions frees up dopamine for further suppression that nurtures the ego, furthers the mind's gaiting capacity, and augers complexity. Thus dopamine suppression is the biochemical equivalent of Freud's repression.

Bottom line is you get the concept that our minds seek to move from situations of energic intensity, excitement, and stimulation to ones of lower energy with entropy lending a hand by drawing the mental apparatus to states of less complexity and organization. We might compare a state of high energy to anxiety. Psychiatrists often find that anxiety and depression are cousins, which is why antidepressants are the best medications to dilute anxious nervousness even though they take weeks to hit home. They will often reduce or even eliminate panic attacks. We all find anxiety unpleasant, and the mind will go out of its way to scuttle it if possible.

How, you might ask, did our minds evolve from primitive and less organized to the energized, precisely constructed thinking machines of modern man burying the primitive organization deep in its watery grave? Surely this requires an increase in excitation and a kick in the face to fiendish entropy's desire, with less organization being the settled state of inorganic matter. What factors could have led to this startling, anti-Freudian apogee? The rather blunt, primitive, wordless, concept-free 1.0 mind of the Neanderthal (or a child) resides at a lower level of overall excitation than the complex, precise, highly differentiate and hierarchically structured mechanism, the conceptual, social, and moral psyche of a modern Homo sapiens. The mental habitat of the child has less excitation than the homogenized machinery of an adult. This clearly goes against Freudian principles, you are saying. You are contradicting yourself, Dr. Lesk, and proposing something that you have now repudiated. The mind could not have gone from a perigee of excitation to a higher one according to you and Dr. Freud. We've caught you in a grave contradiction.

But of course I have an answer for this which is that in its ultimate wisdom, *evolution insisted on it!* How could it not? Once man became a problem-solving, predator-neutralizing Goliath with the invaluable assistance obviously of language and the ego's clever talent in delivering masticated concepts to our prefrontal lobes, natural selection freaked with enthusiasm. The group with the greatest conceptual skills had a solid advantage in surviving over all others, and by the year 10,000 BC we'd decimated all other species of Homo and neutralized most of our predators. This propelled man toward greater and greater conceptual sophistication, mental differentiation, and organization even if it meant a higher state of overall energy and a body blow to entropy. Thus *evolution trumped entropy and Darwin trumped Freud.* Natural selection mandated that survival would trump entropy, and entropy can be defined, for our purposes, as a chaotic, less-organized state of lower energy. We would give up our caveman ways with their simplicity and that trough of mental excitation in exchange for dominance, safety, and security, embracing what I call *antropy*, or anti-entropy, and dopamine suppression. Our minds, with the Oz of language, complexified so to speak and our brains modernized into the intricate, civilized, angst-ridden psyches of modern humans with all the foibles and insecurities tagging along. Clearly children are given this task of complexification, reinforcing executive brain functioning, guided by parenting, learning, reading, discipline, socialization, and so forth. And the trade-off for this leveraged excitation is the safety that our modern civilization with its institutions provides for us. Yes, we have progressed anti-Freudianly and antropically from a paucity of excitation to a uniquely indemnified high point in exchange for the benefits of society. Our minds have advanced from the brutal, terrified, survivalist ethos of Mr. Caveman to the secure plateau of modern-day living. We now expect to survive into our eighties or beyond, to not endure brutal conditions, and to be able to negotiate a society that provides pathways toward success and even happiness, which is one reason I assert that happiness is a modern invention. It is when societies begin to break down and fail in their promises that we begin to question this exchange. If a person loses control and blows away fifty-nine innocent people from the top floor of a Vegas hotel room, we are unable to compute this aberrant behavior. This is not what we bargained for when we navigated the plateau of higher energy that the modern mind embraces in exchange for society's gifts. All of our assumptions and educational activities as we mature rebel against it. We are stunned and reminded of the mind of the panicked caveman with his lower energy and brute primitivity. After all it was evolution and its entropic exegesis that seduced us with this antropic deux ex machina called modern mind 2.0 and its neurotic intricacies in exchange for a society free of the terrifying survivalist nightmare of our Neanderthal forebears.

And so it works for much of the civilized world. We have made the monumental exchange, and humanity has decided to forego at least partly the pleasure principle in its bargain with security, the vanquishing of our ageless predators in defiance of entropy's desire. (I'm not implying that humans aren't pleasure seekers. They are.) Yet that lower state of excitation, brain state 1.0, always lurks in the background. There is always the possibility of a return to an energy trough blessed by Freud's principle and entropy's entreaty, the mind ever capable of resuming its entropic ideals in the playground of inorganic matter. Since I have said that *the primitive always seeks to return*, now we know why and have the rationale behind the primitive organization's ferocity. When it reasserts itself it is gleefully and avariciously transporting the mind back to a lower state of energy and organization in releasing the intensity of gating of our former primitivity that existed for millions of years before language's incongruous arrival. For better or worse the schizophrenic psyche has now returned to a prehistoric's disorganized valley of excitation. Just as the child's mind is in a state of energy vacuum until it matures by suppressing dopamine while embracing language, so is the schizophrenic's when the primitive organization storms the command center with its tsunami of dopamine de-suppression.

So now we know another reason why the primitive organization is so successful, ruthless, and immutable. It is not just that it returns the sufferer to a state of mind entertained for millions of years before the gaslight of language's fortuitous appearance; it has brought the schizophrenic psyche to a lower state of overall energy and gratified entropy with the caveat that the schizophrenic had at one time reached a mature level of functioning that was in blatant contrast with the primitive demands of this mental regression, an anti-primitivity so to speak. (Furthermore, the primitive organization that lies in us all undergoes a plunge into the coffin of the deep unconscious around age five or so. This requires significantly increasing expenditures of energy in defiance of entropy, energy that is released to serve the purpose of the psychosis when the primitive organization makes its avaricious reappearance in schizophrenics.) In so doing, the primitive organization has undone some fifty thousand years of evolution although some of the modern structures are still standing like a church steeple in light of a receding tsunami.[1] In that regard, one might label our blessed antipsychotic medications "evolution pills." They transport the schizophrenic tens of thousands of years forward in time to a place where they used to live before their breakdown. Who knew a little pill called *Thorazine* could accomplish all that?

Looking at the landscape of the evolutionary and anthropological basis of schizophrenia as a whole, Darwin would certainly have had his opinion of this transformative malaise. The cold critique of natural selection would indubitably be on the side of mental development that arose from language with its MIT-like problem-solving ability, its icy conceptual clarity and sophisticated organization. This book forwards the hypothesis that schizophrenia is a product of very recent evolutionary events. Although humans, or something like them, have been around for some six million years, language has been extant for a mere 50,000 to 100,000 years. The onset of language was afforded by skyrocketing advances in brain development, the central nervous system ballooning incrementally over eons of time to the point where the fetus's little head could barely navigate Momma's birth canal. The Homo sapiens brain profited from a unique folding of the cerebral cortex, allowing for greater speed and a sly adeptness at symbol formation that led to speech. In turn, internal language enriched the brain with fertilizing salves called BDNF and VEGF and allowed for the suppression of dopamine with its gating function, enlargement of the ego, and adoption of modern mentality 2.0, all ushered in by the musculature of the mouth, tongue, larynx, teeth, and diaphragm. As we learned to express ourselves motorically, which required dopamine suppression as all muscular movement does (located in the nigrostriatal tract) in a delicate balance with acetylcholine, we shifted this talent for dopamine binding gradually to conceptualization itself, Sherlock's passion, which is located in the mesocortical-mesolimbic highways. Without language giving birth to thinking, we'd be back hurling stones at passing gazelles.

As his deductive talents increased, it became man's highest goal to defeat the tooth marks of the bounding cheetah, the whims of starvation, weather's passion, and the cauldron of disease with which he'd struggled for millions of years and in taming them to neutralize a large part of the crucible of natural selection. Language's panacea led to socialization, agriculture, education, religion, industrialization, and art among other institutions, with conscious thought building all these bulwarks of civilization. Instead of natural selection, sexual selection (survival of the most reproductive) became the guiding principle, leading to a profound evolutionary transformation unique in man's history and in the history of evolution as a whole, which we are still getting used to. But we continue to be on the cusp of that stupendous surge. The momentum of this metamorphosis is with the six-million-year-old pre-verbal brain (or primitive organization) as we have barely distanced ourselves from that primitivity, and at times it finds, in certain individuals who are not yet on board with the program, a means of reassertion. These individuals are designated schizophrenics, or today's dispossessed.

This primitive organization exists in us all. We have succeeded in glossing over it with an ego-driven layer of civilization, the latter representing man's higher aspirations and goals for safety and socialization. On an intrapsychic level we now have a contemplative brain consciousness that allows us to participate in our own thought processes for the very first time. Our cognitivity has matured from childlike to adult with different adult rules. Our speech progresses from idiosyncratic or autistic to universal with the ability to communicate with our fellow humans on common ground. What we have focused in on is that transition from jungle to disco and the trappings of civilization that tag along. A process ensues in which children ripen from primitive to adult, and supported by education and parenting, the vast majority do so quite successfully, enlisting the specific suppression of dopamine first in the arena of movement and coordination but then in the area of contemplation itself. But the old ways grumble in retributive return to the primitive past in a most avaricious manner. Just before some ill-defined point of no return, the *six-million-year-old primitive seeks to reassert itself.* The barge of civilization is shifting and not all of us will shift easily yet. It is rather remarkable that most make the transition admirably with a propulsion of the ego to a heretofore unachievable stability involving dopamine suppression and a rejection of entropy following language's road map. Neanderthals could only dream of attaining these benefits from words, as if language were just waiting to complete us and us alone. As the barge makes its ponderous turn, eventually the ancient ways will finally melt and the new human anti-primitivity will be complete as we distance ourselves from entropy. This will put an end to our caveperson heritage and to mental illness as a whole perhaps some twenty thousand years hence.

We are all absent-minded dreamers, animating our daily machinations with tides of floating ideas. We organize, tightening the kite string of our minds. That state of antropic intensity defies nature's instincts toward disorganization, energy's flight from intensity. Rocks don't think, and water simply flows downhill. But the organic that is ultimately born of inorganic elements dances to the same laws of nature. In a recent shift to intensity, the Homo sapiens mind outdoes everything that dust endures, a machine so refined as to dazzle the entropic domination of the universe. Language leverages a sapiens brain to razor sharpness and folds the carbon atoms to a height unanticipated by Father Time or Mother Nature. We're still in the first lap of the race, turning gradually, some catching up, others fully on board the good ship evolution. Still partly victorious, the old caveman mentality with entropy cheering it on pervades some still, unwilling to forfeit anything physics or matter could endure.

For six million years, man used only primitive forms of communication. Though not decidedly different from that of most species, his brain was his greatest asset. One could describe us as insignificant animals, glorified chimps, in that we were not even at the top of the food chain (Harari 2015). The human brain continued its expansion according to the rules governing all living matter in the universe as Darwin detailed: those mutations that promoted survival would be endorsed by natural selection. "Mammalian brains change their shape by becoming folded as they increase in size" (Hofman 2014). Clearly a larger brain offered advantages and was blessed by natural selection but organization was also key. "The development of the cortex coordinates folding with connectivity in a way that produces a smaller and faster brain."

The forces of adversity in the natural environment were brutal: predators large and small, illness, injury, starvation, and extremes of weather all played a role in the natural selection process. Eons of time allowed these genetic challenges to simmer. Eventually man's brain reached a critical mass and organization. "The relative size of the entire cerebral cortex (including white matter) goes from 40 percent in mice to about 80 percent in humans" (Hofman 2014). The brain of Homo erectus was about 900 cubic centimeters compared with the 1,350 cubic centimeter Homo sapiens brain (Leakey 1994). The use of fire and cooking assisted digestion and therefore a smaller intestine was possible, thus allowing for more energy and blood to be used by the brain (Harari 2015). "The brain accounts for 2–3 percent of the total body weight but it consumes 25 percent of the body's energy" (Harari 2015, 9).

Eventually Homo sapiens and perhaps other hominins became capable of symbol formation, a talent ultimately leading to language. In turn, language enriched the mind according to the rules of developmental psychology with greater specificity, complexity, hierarchical structuring and organization, and, ultimately, thinking. The ego became a word-processing dynamo around which the brain organized and lateralized to the left side, expanding to accommodate all that verbalization was. It became like a sense organ that delivered to consciousness not external stimuli but concepts, allowing introspection as the dorsolateral prefrontal cortex held the concepts up to some grand jumbotron of veracity. There is considerable debate about the onset of language, which may have evolved gradually or, as some contest, exploded onto the scene. Since thinking fertilizes the brain with chemicals like VEGF and BDNF, I believe language likely burst upon the scene. Complexity of artistic expression is presumed to signal the onset of language. "Painting and engraving in rock shelters and caves enters the record abruptly about 35,000 years ago" (Leakey 1994). "After 50,000 years ago, human self expression shifted into high gear" (Zimmer 2007).

Language was the single most momentous development of man's evolution. "Natural selection would therefore have steadily enhanced language capacity" (Leakey 1994). It would eventually upend the sacred principals that guided Darwinian evolution. "The evolution of human language as we know it was a defining point in human prehistory" (Leakey 1994). According to Paul Broca, "the faculty of articulate language holds pride of place. It is this that distinguishes us most clearly from the animals" (Crow 2006). With the help of language and its symphony of blessings, the brutal forces that honed natural selection gradually were blunted. Evolution began a massive course correction that we're still enduring.

Language carried with it the aforementioned lateralization of the brain, enlargement of left side over right as well as handedness and dopamine suppression. "In most people the left hemisphere is larger than the right—a result in part of the packaging of language related machinery there" (Leakey 1994). While chimpanzees do not have this (Crow 2000), some inklings of lateralization were found in Homo habilis almost two million years ago although that does not mean they had full language (Leakey 1994). The ratio of index to fourth finger length may be correlated with lateralization and is lower in Schneiderian schizophrenics (Venkatasubramanian et al. 2011). Conscious thought, impossible without language, became real and quickly proved its importance in problem solving as we learned to speak not just to others but inwardly to ourselves.

Evidence "points to language as the engine of human brain growth" (Leakey 1994). "Various forms of human activity earlier than 100,000 years ago implies a total absence of anything modern humans would recognize as language." "The appearance of modern humans, sometime within the last 200,000 years was accompanied by a tripling of the size of the human brain." "The force that seems to have accelerated our brain's growth is a new kind of stimulant: language, signs, collective memories—all elements of culture" (Leakey 1994). Those species of Homo without language proved to have a dire disadvantage. Language advanced Homo sapiens cognitively to a plateau previously unattained.

Several species of hominin existed simultaneously. It wasn't until thirteen thousand years ago that Homo sapiens stood alone (Harari 2015).

They probably defeated Homo floresiensis, the Naledi, Denisovans, and the Neanderthals who lived 135,000 to 34,000 years ago (Leakey 1994), had larger brains, and were more muscular. "Sapiens replaced all the previous human populations without merging with them" (Harari 2015). At least that is one theory. Other theories suggest there might have been some inbreeding between us and the Neanderthals as we have some of their DNA; alternatively, another account asserts that the other species of Homo and all but one thousand Homo sapiens were wiped out by climatic changes. It wasn't the sapiens' muscularity or even brain size that offered the advantage; it was the ability to speak. It is not unimaginable that the language-advanced Homo sapiens were attacked by the more primitive Neanderthals in jealous land grabs that were occasionally successful even though ultimately the advantages offered by language to us sapiens proved insurmountable. "It may well be that when Sapiens encountered Neanderthals the result was the first and most significant ethnic-cleansing in history" (Harari 2015). The advantages of language proved decisive. "Neanderthals' extinction may have been related to inferior language abilities" (Leakey 1994). The battle of verbal versus pre-verbal man was weighted in favor of the former. "Why did we push all human species into oblivion? Why couldn't even the strong, brainy, cold-proof Neanderthals survive our onslaught? . . . Homo sapiens conquered the world thanks above all to its unique language" (Harari 2015).

One could say that language arose to meet a need as further advancement of brain power by size and configuration alone proved impossible. Nor could a larger skull fit through the female pelvis (Lee and Yoon 2018). "It will be argued that with the evolution of the human brain we have nearly reached the limits of biological intelligence" (Leakey 1994). An intense natural-selective thirst for mental enrichment could only have been met by a paradigmatic shift of language, and under the weight of evolutionary encouragement, necessity became the mother of invention. Just as the need for more and better data organization and processing fostered digitalization, language was, in a sense, a ghostly inevitability.

Here we see a sharp dividing line between pre-verbal and verbal man creating a course correction in the nature and direction of evolution itself. Along with the wholesale transition from natural to sexual selection, it would make sense that a change in predominant neurotransmitter might occur. "Comparison of the human substantia nigra and VTA (ventral tegmental area) with those of other mammals or vertebrates reveals a tremendous increase in the number of dopamine neurons . . . This very large number of dopamine neurons in humans is significant since, in the meantime, the brain size is only two to three times bigger in humans than in Macaca, in close correlation to the body weight of the animal" (Vernier et al. 1993). However, no longer engaged in a savage, daily street brawl for survival, man may have eventually garnered other talents than those afforded by dopamine to propagate his genetic material. The intensity of environmental pressure may have required dopamine while the demands of mental activity and sexual selection may have fostered a renegotiation with the neurotransmitter. Dopamine undergoes a gradual suppression as we mature—perhaps Homo sapiens' greatest accomplishment and that which underlay the dividing line between us and them. This maturational imperative is reinforced by education, child rearing, socialization, laws, religion, jurisprudence, and so forth. In Parkinson's disease, dopaminergic suppression may be excessive leading to late dopamine cell death. The tsunami of dopaminergic tone (de-suppression) in schizophrenics allows us to treat them with dopamine-blocking agents. In Alzheimer's disease what's lacking is not dopamine but its partner in dance, acetylcholine.

In suppressing dopamine the old brain borrowed a trick from musculature. A balancing act akin to highwire stunts, dopamine needed to fall into line with another neurotransmitter, acetylcholine, to grease the wheel of coordinated movements. The two chemicals dance in unison once dopamine is reined in. Mouth, lips, and tongue, used mostly for food processing before language, now took on a delicate communicative role: speech, and this coordinated enunciation too required dopamine's humble submission. As we adopted the pleasure of pensive enjoyment, learning to self talk, which requires subvocalization even if we're not speaking out loud, this tangible talent for dopamine suppression fell from the motor tract inward to the previously untouched mesocortical-mesolimbic dopamine tracts. (Tourette's syndrome and OCD are prototypical illnesses that highlight the transition from involuntary tic to semi-voluntary compulsion with the addition of language to grease the wheel.

They are often seen in illnesses like schizophrenia and autism.) For the first time a brainy animal could suppress dopamine in a tract other than one related to muscular coordination (the nigrostriatal), a tract purely contemplative like the mesolimbic. Thus contemplation too was in line with dopamine's suppressive harnessing and it was, for the first time ever, utilized to promote brainy activities: thinking, contemplation, deductive reasoning. As it danced toward inner speech but maintained a subvocal basis, thinking borrowed the delectable habit of dopamine suppression from its friend the motor system.

Both Parkinson's disease and schizophrenia (and many other illnesses) may be a result of evolution's course correction and our renegotiated relationship to dopamine. In relation to Parkinson's disease (which only occurs in humans as does schizophrenia; Crow 2006), a late-life dopaminergic aberration, Vernier et al. (2004) states, "What has been selected upon specific functional constraints may not be so beneficial several thousands of years later, when the species do not live in the same environment . . . Many human diseases are very often the consequence, the outcome of an evolutionary process of character selection which was made for another purpose and in another context than the one in which humans are living today." Darwin would have agreed, I assume, that the rules changed after language's fortuitous appearance. Natural selection's decision was a no-brainer so to speak. The brain of modern sapiens with their more intense cognitivity, hierarchically structured mind, and conceptual consciousness trumped Mr. Neanderthal's primitive brutality. Homo sapiens proved that. But the primitive organization's old ways are still trying to recapture territory, the territory between our ears.

We're barely in the inception of this very recent, in evolutionary terms, and profound course correction as evolution abandons natural selection for sexual selection. Inevitably there are still some, given this revisionist dogma, in whom the old ways succeed in reasserting themselves. This primitive organization appears to strike randomly although genetics may embrace a general propensity for dopamine de-suppression as a whole, leaving behind a scattershot of mixed diagnoses and neurological debilities. (If we look at polygenic risk score, or degrees of similarity to the typical gene arrangement of an illness, we see not just schizophrenia but major depression, bipolar, anxiety disorders, and others increasing in odds as well, hinting tentatively at a global genetic risk for the trait of dopamine de-suppression as a whole. This makes perfect sense since dopamine de-suppression is at the bottom of each mental illness we treat.) We find what are called *soft neurological signs* in relatives of schizophrenics, subtle differences in their brain functioning, as if evolution takes a swipe at a whole family, only one of whom becomes schizophrenic. So for example if one inherits a mega-vulnerability to de-suppress dopamine, one has a higher likelihood of developing any one of a number of mental infirmities. If one inherits a low propensity for dopamine de-suppression, then one is relatively protected from mental illnesses except under extreme environmental circumstances. There may be genetic and experiential vulnerabilities that interact in a grand symphony. Since there was possibly crossbreeding between sapiens and Neanderthals, one might surmise that schizophrenics have more genetic material from our extinct ancestors, but this has not been born out by research.

If we look at the onset and course of schizophrenia, we see the footprint of the primitive organization all over it. Analogous to a Neanderthal attack on the more advanced sapiens, the theater of battle has shifted from the steamy jungle to the context of the human mind. The brain matures en masse, both the primitive and modern areas, and as maturity is reached, the primitive organization is ready to mount its unforgiving assault on the territory of the modern mind and proceeds in a fairly avaricious insurgence. The suppression of dopamine is not yet fixed in the chosen individuals. (There may be a point of no return past which the primitive organization is no longer able to undo the suppression of dopamine that has hardened during maturation.) Hallucinogens reverse this dopaminergic suppression chemically, and some illnesses may pry it from its mooring, resulting in psychosis.

Some schizophrenics may experience on-the-way-to-psychosis phenomena. The budding schizophrenic realizes that they are being attacked by some radical and inexorable demon and reacts with depression, drug use, desperately random help-seeking behavior, or atypical symptoms as the dopamine de-suppression makes its way through various tributary options. Schreber was hospitalized with hypochondriasis eight years before his dramatic breakdown. Elyn Saks suffered a major depression before the onset of her schizophrenia. So the de-suppression explores various pathways forward like an army staging brief forays into enemy territory. Once the primitive organization succeeds in this ruthless takeover (which it does in 1 percent of the below-age-twenty-five population), the higher centers of the brain are left underutilized or short-circuited and are prone to atrophy. They self-stimulate as they do during sleep, resulting in hallucinations. It makes sense that the latest evolutionary attainments, promulgated by their captain, the ego, might be the most high-profile targets (Crow 2006). This atrophy may be so severe that antipsychotics no longer do much to improve the outlook, the adult brain being left crippled, unable to regain its former glory. The illness foments in stages, not a static diathesis, a several-year process during which the primitive organization degrades and diminishes the ego in a battle royale, the voices taking on a decidedly pejorative tone. Over time, positive symptoms give way to less-dramatic negative symptoms as entropy enjoys its final victory, a victory unequaled in the universe as our caveman could never be so entropically compromised. Language's decisive victory over entropy's desire sets up a situation that when it reverses allows the inexorable entropic seduction to drag the schizophrenic to the lowest energy imaginable. In Schreber's case, this coincided with his abandonment of all resistance to his delusion of being a woman.

"Psychosis," or roughly "craziness," needs redefinition since it's a term tossed around loosely in various situations of mental compromise. It exists always in the context of cognitive regression, a return to the childlike mindset 1.0 of our Neanderthal brethren before language. Schizophrenics are adults repossessed by a primitive mental rulebook from our prehistoric, prelinguistic past. Psychosis takes place in the context of this cognitive backslide. "Psychology of the individual and of the human race, animal and child psychology, psychopathology and the pathology of special states of consciousness—all these can be approached from the genetic standpoint" (Werner 1948). By genetic, Werner means developmental, or the progression of each individual from child to adult. "The question now arises whether the pathological mentality can be integrated within a theory of mental development. Can the 'childishness' of many . . . (schizophrenics) let us say, be related to normal childhood, or the primitiveness of so-called uncivilized peoples?" Children can be characterized at times as daffy, crazy, psychotic in word and deed. Younger kids are in the early stages of language development in which words are used uniquely, on their own terms. "The functional representation in the strict sense which expresses itself in the capacity for communicating a cognition by symbolic formation (gesture, sound, writing, drawing) moves through a course of development from a syncretic (implicit) symbolism to one that is pure and detached" (Werner 1948). Their responses are idiosyncratic, not universal, and are not understandable except in relation to the child's universe. Children see language in relation to action and in combinations of actions rather than a crystal-clear denotation of concepts universally understood.

Schizophrenics (and hallucinogen users) dwell in that same haggard mentality, their thinking more diffuse, magical, and lacking adult rules of logic. They too express things idiosyncratically with reference to their own uniqueness, which may not be so easily decoded by others. The fluid boundaries between themselves and the world around them blur, reflective of ego weakness and the lack of the suppression of dopamine.[2] Primitive people display a similar mode of pre-logic. For example, in some cultures all windows and doors are kept open if someone is in labor to facilitate the birth. "The development from a diffuse perceptual organization characterized primarily by 'qualities of the whole' into an organization in which the essential feature is a decisiveness of parts standing in clear relationship can be observed in the growth of the child's mentality" (Werner 1948). What does that mean? Schizophrenic thinking (like children's) plays by quirky, more-diffuse rules. Gradually as one matures, the rules change in favor of the communicative clarity and differentiation of adult logic, placing words at more of a distance, now tools for daily communication. But for the schizophrenic, just when maturity comes calling, the primitive ways bubble up from our past, and the three yardsticks that represent six million years of hominin existence come back to haunt 1 percent of us in that final one inch.

Now that we've defined what happens in schizophrenics and exactly where it comes from, we are no longer in the dark about whom we are treating and what is happening to them. By describing schizophrenia in this way we can confront our patients with understanding, not ignorance, and we can account for or at least begin to define much of the hardcore data on schizophrenia as a whole. Let's review.

The central paradox of schizophrenia (Crow 2000) questions why this ancient affliction is not extinct. According to Darwin, an illness that starts early in the reproductive years, reducing one's likelihood of survival and production of offspring, is significantly unlikely to persist. The fecundity ratio, or reproductive rate, of schizophrenics compared to non-schizophrenics is much lower. However, according to primitive organization theory, schizophrenia is an evolutionary event, not a genetic one (although there may be inherited vulnerabilities to the process of mental illness as a whole related to a propensity for dopamine de-suppression) falling decisively outside Darwin's rules of evolution, and on that basis it will persist for thousands of years more until the course correction of evolution is complete. When this happens we'll be much farther from the lingering primitivity of our past and greedy entropy's seductive force. Despite intense scrutiny, no genetic etiology or transmission of schizophrenia has ever been identified, only clusters of small genetic errors (single nucleotide polymorphisms) that coalesce not only around schizophrenia but the other major mental illnesses as well. What could be inherited in fact is a propensity for dopamine de-suppression, exposing one to the potential for a host of different illnesses.

The age of onset is governed by the fact that not until the primitive organization that seeks to storm the bulwark of our mind biologically matures can it launch its invasion of the modern 2.0 structures. There also may be a time before dopamine is fully suppressed that the window of opportunity yields exposure to attack. In addition the weight of the repressive material swells over time requiring greater and greater energy, which, for the vulnerable individual, may reach a breaking point.

Both cortical and subcortical deterioration (Stabenau and Pollin 1993; Moser et al. 2017; Cahn et al. 2002) with ventricular (brain space) enlargement, and frontal and hippocampal desaturation is seen in schizophrenics. "Cortical thickness measures most highly correlated with the non imaging variate were widely distributed in the frontal, insular, temporal, parietal, and visual cortex" (Moser et al. 2017). According to primitive organization theory, degeneration is due to their diminishment of function now short-circuited by a more ancient, cognitively less-developed entity. The modern brain takes this as a state of sleep and begins daylight dreaming or hallucinating to ward off the dangers of atrophy but is clearly not always victorious in this endeavor.

This short-circuiting has some ramifications: hallucinations as in Charles Bonnet syndrome with the executive branch of the brain government self-stimulating to remain active and viable, inflammation of these potentially decaying tissues, and a battle royale between a rampaging, primitive 1.0 mind and the remnants of the modern 2.0 edifice.[3] As the disease progresses there is greater decay of neurons with resultant tissue inflammation and the gradual loss of traction of antipsychotics over time. This degenerative process can be influenced by early treatment and the actions of the patient in regard to utilization of their higher cortical functions (thinking, studying, reading, brain exercises). Cognitive remediation, a newer method of restoration, succeeds by engaging executive centers of functioning.

The success of dopamine-blocking agents, the Thorazines and Haldols, can be surmised to be due to the high dopaminergic load in primitive humans (more prevalent in humans than in other species) in contrast to the intense suppression of dopamine over the last fifty thousand years of evolution. This suppression may be less complete in schizophrenics or more intense in Parkinson's disease patients although these issues remain unclear. Whatever the case, when the primitive structure stages its dramatic comeback, it does so with a dopaminergic intensity that can be blocked by antipsychotic medications, which put a palliative halt to the invasion and restore the gating function.

If schizophrenics (especially those with Schneiderian symptoms) have lower second digit to fourth digit length ratios, this presumably reflects the degree of lateralization in the brain (although it also reflects exposure to androgens in utero as well), a shifting imbalance intracranially that arose to accommodate language's multiplicity of facets and accoutrements. We might predict lateralization to be less robust in schizophrenics as this may signal some deficit in "language-ness" and corresponding ego weakness that, as long as we're entertaining deep speculation, could reflect a vulnerability to attack by the primitive organization. Autistics too have language deficits with the severity of illness corresponding to the greatest linguistic weakness.

Pre-pulse inhibition (the reduction in startle reaction to a loud noise by a dull pre-pulse sound given milliseconds before) is inhibited in schizophrenics, children, and those taking LSD. This can be simply explained by the diffuse nature of their mind's attentional locus, a de-suppressive haze lacking the gating function necessary for clarity. With a more-diffuse attention span, less is captured by the dull thud leaving more to respond to the gunshot pulse (with less gating of it). It's no surprise that the failure of pre-pulse inhibition can be blocked by antipsychotics. Interestingly, children's failure of pre-pulse inhibition gradually diminishes as they mature and suppress dopamine, giving clarity to the brain space: the gating function in action.

The core symptom of schizophrenia is then a wholesale reversion to the primitive organization, differentiating it decisively from bi- or unipolar disorders of mood that reinhabit a highly functional plateau between episodes. It is not a dementia, but its victims certainly experience a regression to a more primitive, prehistoric, and prelinguistic mode of thinking that was here for some six million years or so. Complicating the picture even further, schizophrenics are subject to atrophy of frontal lobe areas like the dorsolateral prefrontal cortex simply because these are no longer being used as much, having been replaced by a different center of operations. They try to make themselves useful to avoid the dreaded cell death of atrophy by the self-stimulation of hallucinations, which at night are called *dreams*.

Why certain individuals are afflicted with this takeover by the primitive organization and others are not is unclear, but our proximity to this recent rejection of primitivity, still in entropy's orb, makes us vulnerable enough to recapture that 1 percent of the population. One clue, lateralization, reflected in the 2D:4D ratio, may be less robust in Schneiderian schizophrenics (those suffering from the more definitive Schneiderian symptoms [Venkatasubramanian et al. 2011]). Since lateralization is indicative of language acquisition and its adjuncts, one might attribute this to a failure of language-ness or enculturation. However, there may be a singular demon of randomness in the affliction, an evolutionary glitch that strikes without cause. Perhaps there is some inherited global vulnerability to mental illness resulting in a multiplicity of potential diagnoses, perhaps a variable propensity to de-suppress dopamine, in which case individuals high on the scale are more vulnerable to external stresses, which may push them toward one psychiatric affliction or another. The backflow of dopamine has many tributaries it can follow leading to different symptom profiles. This would be an explanation for the interaction of environmental stress and genes. But some illnesses simply occur on a random basis, the barge of evolution having recently turned and not everyone yet able to turn with it.

The specific anatomical structures involved with the primitive organization have yet to be defined. Clearly the more primitive parts of the brain are responsible, but the primitive sits there in all of us not far from our modern sensibility, which is layered over it. We have access to it merely by slugging some bitter ayahuasca tea or when the brain is compromised as in Alzheimer's or when we're sensorily deprived in a floatation tank or in the entropic nirvana of sleep.

Even so the interaction of the primitive organization with Freud's psychodynamic framework isn't all that clear. The ego (the crown jewel of the last fifty thousand years of evolution), which stands in opposition to primitivity, certainly suffers with dopamine de-suppression and thrives in the suppressive think tank. Along with it, reality testing and prioritization of incoming stimuli are critically compromised when gating is lost with dopamine de-suppression under the influence of entropy's seductive tune. Symptom selection is no doubt a product of the individual's personal psychodynamics as in Schreber's paranoia. Could it be then that what Freud was really driving at in his grand mechanistic and therapeutic creation was a furthering of the neurochemical shift to dopamine suppression?

The profound shift in evolution from natural selection to sexual selection and modernity may have solved multiple problems, bringing us to a higher plateau of mental power and organization, but it also created others as we renegotiated our complex relationship with our wily friend dopamine. Schizophrenia, Parkinson's disease, autism, suicide, diabetes, stuttering, Tourette's syndrome, ADD, and Alzheimer's (to name a few) are potentially rooted in evolution and inflicted by our changing dialogue with dopamine.

At the end of his analysis of Schreber's memoirs, Freud says, "And I am of the opinion that the time will soon be ripe for us to make an extension of a principle of which the truth has long been recognized by psychoanalysis, and to complete what has hitherto had only an individual and ontogenetic application by the addition of its anthropological and phylogenetically conceived counterpart. "In dreams and in neuroses . . . we come once more upon the child and the peculiarities which characterize his modes of thought and his emotional life." "And we come upon the savage too [thus we may complete our proposition] upon the primitive man, as he stands revealed to us in the light of the researches of archeology and ethnology" (Freud 1963). I believe Freud was admitting that it wasn't enough to compare mental symptoms to the illogicality of children but to crawl further into our chaotic past to a time before language to comprehend completely the effects of time on our weary souls.

In order to understand who we are and who we are becoming, we must countenance our lengthy past and the radical modulation of our fortunes wrought primarily through the acquisition of language. We are still in the midst of that glittering evolutionary overhaul, but ten thousand, twenty thousand years from now we may not be, eliminating schizophrenia and mental illness as we know it from our repertoire. Only time will tell.

The winds of evolutionary change swept over us swifter than any hurricane or sea squall. Like a tortoise suddenly placed in an elevator or a giraffe stowed on a yacht, we found ourselves parachuted into uncharted territory so we struggled to catch up and realign our former mental simplicity with the odd streets of this new world. For the most part our midnight cowboy learns to survive, but at times, overwhelmed by the foreign nature of his revisionist mission, he collapses in confusion, harkening to less-complex times. Hit with a global change in assignment, our minds revert to disquietude, angling back to familiar territory. With entropy chuckling at its victory, some remain imprisoned by the past, sucked back to what was.

It's important to note that this is a whole new model of disease creation. A swing in evolution that is only partly realized leaves behind those who have yet to fall under its wing. We need to examine this point of view for clues to other maladies as well. Not only did evolution change, but our minds did as well, enriched in a paradigm of intensification and complexity against the second law of thermodynamics, entropy, leading to our ability, for the first time, to participate in our own thinking. This earthshaking development created the need for our minds to make dramatic alterations in functioning.

This changed the world in that we used our minds to conquer predators, to create new environments, new levels of physical activity, socialization, cooperation, and more. How these changes benefited and possibly harmed us mentally and physically deserves more intense scrutiny. Humans went from experiential and physical beings to conceptual ones, and one could surmise that in the future we will become even more brainy still. The changes in sedentary lifestyle alone are staggering. Dietary changes might have led to diabetes since there may be different levels of pancreatic reserve. The explosion of carbohydrate intake that moderns indulge in may surpass the limit of the pancreas to endure, resulting in either childhood diabetes or later onset type 2 diabetes. We must be careful not to outsmart ourselves and in vanquishing the predators that plagued us for millions of years create new ones. Having moved from chaos to order, we need to appreciate order's value, to protect and enhance it. Any slide into chaos may well be swift and irreversible.

* * *

Let's go back to our time tower. This time we'll add future vision as one of its properties. We climb the tower and look out over the horizon and see our future looming ten thousand or more years ahead. Assuming mankind is still on earth (a bold assumption at that), we might very well see a small person who is more conceptual and less physical than we are. We will probably see that person using a higher form of logic than is customary in modern society currently. What this form of logic looks like is unclear. Undoubtedly we'll observe a person who is less experiential and more conceptual than we are. (This raises the question as to whether logic has a continuum of advancement or has reached a plateau with the leap engendered by language. My guess, obviously, is the former.) We'll see a human studying ancient life in the 2000s and shaking his head at how primitive we were, as primitive as we regard Mr. Neanderthal. This new human may well be a highly brainy character but short, using his tiny body minimally. Digital or some other processing paradigm will have progressed to the point where this person will merely have to say something (or perhaps just think something) and it will show up at his door in seconds. The tiny humans may never have to leave their houses. From our tower perch we see them living under giant domes of climate-controlled atmosphere. We can view robots doing the physical labor, bringing our futureperson anything they desire. Evolution's course will have fully delivered us to a new frontier. From the jungle we will have arrived at a disco so unimaginably refined as to be barely conceivable today.

And what of schizophrenia? What of mental illness in general? As we look out over the mystical future horizon, we see humans doing things in a very specific and hierarchically organized manner. Their brains will be so refined as to be impervious to their past. Remnants of their ancient primitivity will be so far removed from them as to be untenable. The mental ashes of their primitive ancestry will no longer be holding them hostage, and entropy will have been reduced to a whimpering sow. That primitive organization that is in all of us will be many layers removed from this futureperson's access. They will need stronger and stronger chemicals to trigger our own primitive mental ancestry where today it sits just below the surface and available for our enlistment or, for some, its resurgence. The suppression of dopamine will be increasingly intense. As I look across that future landscape, I see no mental institutions. Psychiatrists are used to modify one's experience slightly in one direction or another. I see in one office a psychiatrist writing a prescription to help one futureperson become more musical, another more inspired, another more euphoric although some of this will be done genetically. I can see futureperson living within his mind, relying on his physicality minimally, conceptualizing endlessly. Births will be planned intensively, conceived outside the body, and probably nurtured from embryo to delivery in a mechanical device. Women may lose their ability to carry children.

Sexuality will be a pastime, not a reproductive endeavor unless one chooses to do it the old-fashioned way. Or there may be humans bred just for their sexual attractiveness to be used for sexual pleasure.

Once the barge of evolution has shifted, what will determine its direction going forward? Whatever cuts into one's chances of reproducing will be an evolutionary loser of course. It's inevitable that sperm and ova will be refined by DNA manipulation, allowing humans to choose (within perhaps certain limits) their child's characteristics. Cunning and brain power still hold an advantage, and this presumably will be true in the murky future as bodily characteristics shrink in importance. There may well be strict controls over the DNA allowed to be conceived, and this will extract evolution's influence considerably. Once reproductive results become entirely under our control, we've removed evolution's symphonic spell. It will be in free fall. Evolution requires a significant force of adversity to prune the myriad mutations that occur spontaneously. With humanity's complete control over mutational occurrence (we'll be able to fix some and create others) and the diminishment of life's perilous circumstances, it's difficult to predict what impact evolutionary forces might have if any. Humans may live two hundred years or more. This will require not just bodily health but a way to preserve the function of the brain for that lengthy period of time. Births will have to be strictly controlled since this longevity will soon use up the earth's capacity to house and protect mankind.

If mighty evolution shifts into oblivion, where will we head in our genetic advancement? Once we remove all elements of chance (e.g., mutations, illness, suicide, and accidents, which will be much fewer) evolution will go into decline as some people believe it is now. Our equation protoplasm + mutation x time = man cannot operate with the thorny adverse forces of mutational pruning removed. We will have replaced natural selection and sexual selection with our own decision making. At that point, however, will adversity find ways with which to impose itself? Having taking over the selection aspect of evolution, will we have inadvertently created a resurgence of uncontrollable forces via global warming or other unforeseen nightmares? At that point society may break down with a return to the entropic chaos that was mostly overcome when Mr.

Neanderthal left the earth. The bargain humans made to increase mental excitation in exchange for safety could conceivably reverse itself, leading to the stuff of apocalyptic fiction.

These are all speculations, interesting as they may be. Humans may find a way to turn global warming into an opportunity rather than an affliction. Other adversities such as deadly viruses or asteroids, to name a few possibilities, could become routine occurrences with routine solutions. But will the greatest predator of all learn to cooperate? Will we find a way to tame the most cunning of all adversaries and live in unison? I speak of course of humankind.

As we look out of our time tower, we see far into the future. What I see is a world of great beauty, potential, progress, power, and harmony. The world is united by digital communication into one entity, and humans have learned (perhaps after some tragedy like a nuclear holocaust) that we are all on this fragile spaceship earth together. I see one global government working to help all people, the blue planet operating in a streamlined fashion, meeting the needs of all. This is utopian perhaps, but humankind will gain more of the skills needed to accommodate global needs over time, and horrid necessity may be the ultimate mother of invention. At some point there will be the realization that strife and discord simply help no one, that cooperation is key, and that we are more alike than different. I see religious beliefs taking a backseat so that they no longer separate us but uplift us with meaning. The most valued possessions will be beauty of mind, conceptualization, and human advancement. There will be no need for guns. Violence will be something shown in a museum. Starvation will no longer be an issue because it was solved eons ago. Hospitals will be merely for advanced tune-ups to aid procession into the years ahead. And schizophrenia, and all mental illnesses, will be a thing of the past, studied in digital medical history as a curious though painful evolutionary chapter.

Part III

Chapter 8

The Polars Uni- and Bi-, or the Invention of Happiness

"How does our primitive organization theory apply to the vastly larger numbers of people who suffer from affective (mood) disorders?" you ask. Let's take a look at some facts about these serious illnesses.

The numbers. Some 15 to 20 percent of the population will suffer a major depression (or unipolar illness) in their lifetime, and they usually recur, meaning it's rare to get just one episode. Two-thirds of the sufferers will be women—why, we don't know. One to two percent of the population will have manic depression, an older name for bipolar disorder. That's nearly twice as great an incidence as schizophrenia. Unipolar sufferers only go in one direction, which means they just get depressed. Bipolars have the lucky distinction to get depressed *and* manic although depression is more common. In short they swing to both ends of the affective spectrum. But the manic episodes can be prolonged and dramatic or short and not as dramatic. We call people who suffer from the latter *bipolar 2*. They are less likely to have the cardinal features of classic mania like grandiosity. Grandiosity means, as a symptom, that people start believing that they can do all sorts of incredible things, like, uh, write a best seller about schizophrenia, produce a hit movie, and so forth. Sometimes this becomes delusional and the manic starts spouting things that are clearly out of touch with reality's hoary grasp like "God speaks only to me and tells me how to cure the world" or "I can control the weather" or "I can just think about something and make it happen." These are what we call *mood-congruent delusions*. They go along with the major thrust of the person's euphoric mood or alternatively depressive state.

We are talking about people whose moods shift deeply enough to warrant a disease label. This is not some weekend blues or relationship-breakup sadness or a grief reaction to the death of your pet dog Sparkler (neat name). This is the kind of mood change that affects almost every aspect of one's life. Numerous qualities of their existence are affected in an episode of mania or major depression. When bipolars get depressed, they usually go into a major depressive episode, and it's often as they are coming down from a manic episode.

Both poles suffer serious sleep disturbance. In short, they can't sleep soundly at all. Early morning awakening is a common variant of the depressed person's insomnia. Depression is classically worse in the morning, classically worse in winter. What we have called *atypical depression* reverses that. They also have difficulty falling asleep, and they have what is known as middle insomnia, meaning they wake up a lot. It's unusual for someone to be in a major depression and sleeping well. Sleep may become the primary focus of their treatment. It is safe to say that until they *are* sleeping well, they probably aren't well enough.

Manics have a strange relationship with sleep. To be blunt about it, they don't. Manics can go for days with no sleep. (Rats die if they are kept awake for two weeks.) One could say they are too ebullient to sleep. They may be too energized to slow down and sleep. Or they may be using sleep to fuel their manic furnace, which they enjoy. Sleep is simply irrelevant to the manic. They are up and running at a thousand miles an hour. Who would want to sleep under those circumstances? (One must carefully inquire if the manic episodes happen when they are using drugs, like stimulants. If so, the episodes don't count diagnostically as mania although, of course, manics can use drugs as well to fuel their manic furnace.)

Then there's seasonality. When I worked in New York I noticed a seasonal influx of mood-disordered patients in fall and then early spring. But when I moved to Minnesota it was much more pronounced. Why? Higher latitude with greater seasonal variability and a more-dramatic change in light hours. The sun starts to lose its potency earlier due to

Minnesota's distance from the equator in relationship to the sun. May is the highest suicide month, which seems illogical. As the light increases, moods should improve compared to the dark expanse of winter. But it could be (my theory) that as the predictable depression of winter starts to dislodge and things fall apart with sleep, concentration, energy, mood, and so forth, people are reminded that they are ill. They then lose control symptomatically and become hopeless and suicidal. Clearly daylight hours have something to do with mood. Lest we forget, humans evolved in warm climates with plenty of sun. Would it surprise you that the farther one is from the equator, the greater the suicide rate? Our eyes are sensitive to the diminishment of light and hours of daylight. We didn't have bulbs or even candles for millions of years (although we had fire, but one cannot stay up forever tending a fire) and our ancestors slept by the seduction of dusk. We didn't have artificial light until relatively recently. We do use full-spectrum light to broaden our winter daylight hours with some success. People with light boxes can get up early and sit in front of bulbs that mimic the sun's wavelength. It helps.

Classically depression is worse in winter and in the morning. There are many exceptions. Difficult anniversaries such as memories of conflicted holidays may trigger a major depression or the anniversary of a divorce or a death.

Not surprisingly, appetite dies with depression. However, some people, those atypical depressives, overeat.

"And what about concentration?" you ask. For unipolars (depressives), concentration is classically horrendous. A student who suffers from a major depression has a very hard time doing their work. They can't sleep and it is nearly impossible to focus. Imagine trying to get through college under those circumstances. Concentration is a function of what we called (and Freud called) secondary process thinking. We drew this distinction earlier. Just to recap, evolutionarily newer secondary process thinking is the kind of thinking that is participatory. It is deliberate contemplation, deductive reasoning. Primary process thinking is on autopilot. It is ruminative, rambling daydreaming that is largely controlled by the brain itself. Depressives ruminate a lot. They can't get back on track. Secondary process thinking is the product of language, obviously, and the ego's ability to feed concepts to the higher centers of the brain. We discussed how, when the ego does this, the brain expands through arborization and connectivity with the help of neurochemicals like BDNF and VEGF. The antidepressant ketamine, which is now the new wonder drug for depression, also stimulates production of BDNF and VEGF. Would it be safe to say that secondary process thinking has antidepressant and mind-expanding properties? After I broke my hip (don't ask me how I got hit by my own car), I was given ketamine preoperatively. My wife and daughter said they'd never heard me talk so much. I was, to say the least, euphoric.

So the loss of secondary process thinking is a depressive event. Is my thesis then that thinking is an antidepressant event? Yes. People who think a lot do have a certain edge of euphoria. You can see this on college campuses, although perhaps you just attribute it to the animated idealism of youth. Active brains are happy, even giddy brains. Depression is accompanied by the frustrating inability to do this, or a withdrawal from secondary process thinking. It is also my theory that what interferes with this ability is the loss of dopaminergic suppression. You remember that language set up a paradigm of normal functioning in which the ego feeds conceptual stimuli to the forebrain. To do this it has to suppress dopamine to clear away the brain space that allows concentration and promotes gating of unwanted stimuli, expanding the ego's power. If dopaminergic suppression is lost, so is concentration, and, so my theory goes, dopamine is de-suppressed as a feature of all mental illnesses.

Manics, in their effort to avoid depression (which they dread), think a lot. They try unsuccessfully to think their way out of depression. But that thinking soon loses organization and flies off into pressured speech that is rapid and difficult to interrupt. The next level, called *flight of ideas*, is a staccato, flimsily connected avalanche of verbiage. Clearly there is a loss of ego influence there and loss of dopamine suppression with its gating function. When they're manic, bipolar patients commonly report racing thoughts, thoughts going so fast as to be impossible to grasp. Clearly the autopilot aspect of their minds has taken control and the loss of dopaminergic suppression has taken a more intense turn. With depressives, the ruminative nature of their thinking harkens back to the autopilot mentality of Mr. Neanderthal although now with the addition of words. It's as if words were added to the primitive thought process and are appendages to it not shapers of it. We have then two different kinds of primary process thinking: the ruminative kind in depressives and the racing kind in manics. Both display loss of ego control and loss of dopaminergic suppression and and harken back to a pre-verbal mentality when concentration was, in essence, irrelevant.

Let's analyze these phenomena in a neurochemical model. As you recall, the process of maturation involves suppression of dopamine (my theory). Pre-verbal humans had plenty of dopamine, which helped them survive, compete in combat, and flourish in overall acceptance of the daily challenges that plagued their lives before civilization. Natural selection gave it a thumb, even two thumbs, up.

One can speak to the necessity of fear. Animals would be in dire straights without fear. If a charging bull elicited no fear, the outcome would be disastrous. Every animal has a fear response. The flight-or-fight reaction is moderated by the adrenals, small glands that sit on top of your kidneys. In an emergency they secrete norepinephrine and the individual takes a flight-or-fight response to the oncoming adversity. Or if something awful has happened, the animal needs to know that. It signals this by pulling back on neurotransmitters in the brain. The mechanism for this is probably what we call *autoreceptors*. These cellular thermostats sit on the nerve terminal and send a signal when there is too much neurotransmitter in the fluid between nerve cells. This signal tells the cells to stop producing neurotransmitters. These autoreceptors are the likely candidates to signal to the brain to stop producing neurotransmitters across the board: serotonin, norepinephrine (yes, the same one secreted by the adrenal glands but in the brain), and our old friend dopamine. This conserves energy, reduces appetite, and puts the person in a state of hibernation such as animals use in winter to conserve energy and stay put. This state might also serve animals well who perceive a nearby predator and want to stay perfectly still until it has moved on. Respiration slows and peristalsis (gut churning) is reduced as the animal tries to blend into their environment.

But words changed the dopaminergic landscape. The ego learned to scoop out words from the preconscious and deliver them to the higher centers of the brain in the form of organized, syntactical thoughts through the auditory sensory channels. When this happened, dopamine was suppressed. Probably one of mankind's greatest feats was the suppression of dopamine. When we think, there is a resistance to it, a strenuousness, at least in the deliberate, secondary process type of thinking. Is that resistance the effort of dopamine suppression? I believe so *and* the effort to overcome entropy. There's a global dopamine suppression that takes place as we mature and a further suppressive push as we're thinking. This could be what separated the sapiens species from others: the ability to suppress dopamine in the central tracts. But if the species sapiens is able to suppress dopamine effectively and at will, perhaps this is its greatest accomplishment. The brain of Homo sapiens took to language like no other and in a way that changed its mental functioning. Mr. Neanderthal or Denisovan was presumably unable to do this. It is not impossible that language created Homo sapiens. How? Language set up a competitive demand. As words increased in use, those who could master them best took on a decided natural selection advantage. This brain-word interaction favored a cerebral organization that could use language in a transformative way. Eventually the brain that used language as a metamorphosis was born and accelerated. It hit a warp speed with us humans. Thinking fertilizes the mind with BDNF and VEGF and our ego strengthens, our divisions between self and world strengthen, our awareness of death increases. Dopamine suppression results in gating out unwanted stimuli, a function of the need for mental clarity that collapses when schizophrenics break down in a swirl of unleashed dopamine crashing back to a 1.0-style of thinking. And the prototype of this suppression came from the motor system, where nigrostriatal dopamine needs to be suppressed to remain in balance with acetylcholine and promote smooth, continuous movements. When this is lost, as in Parkinson's disease, movement is jerky, tremulous, shuffling.

We can compare this evolutionarily to lactose intolerance. Lactose intolerance used to be the norm. Only infants were lactose tolerant to facilitate their suckling at the mother's breast. But once we became a dairy and farming species relying on milk as a staple of our diet, lactose intolerance became a hazardous drawback. Mutations that lengthened the years of lactose tolerance gained natural selection's thumbs up. Gradually this increased to where most of us are lactose tolerant well into later ages. In short, a new demand created a mutational blessing or gradient by evolution's Siskel and Ebert. I might also compare language's effect on hominins to the intensity of the hallucinogen venom of the Sonoran desert toad. Let's consider the following analogy.

At birth we're swimming in a stream (of dopamine). As we mature we gradually learn to crawl out of the rippling river by lifting ourselves up using a branch of a tree overhanging it. We pull ourselves out by suppressing dopamine although parts of our body like our legs may still be in the water. When we think, we pull ourselves out of the stream completely. High and dry, we're able to contemplate ideas in a free, cognitive manner. We're above the impact of rushing water. The wily dopaminergic stream is always there to suck us back in, and at rest, so to speak, we are still at the least dangling our feet in it. This represents primary process thinking as opposed to secondary process where we are above the waterline. Pre-verbal man swam in those dopaminergic waters. When the primitive organization reasserts itself in schizophrenics, the waters are raging and the person is sucked in completely, bobbing breathlessly under the surface. He lets go of the branch under the weight of the raging water and is engulfed by it. Antipsychotics reduce the intensity of that raging water, allowing the individual to reach out and grab a branch, pulling themselves out of the dopaminergic morass. Yet not every schizophrenic will take advantage of that opportunity because it requires cognitive effort, studiousness, or, as in the case of Schreber, someone who is going to reflect mentally on everything that is happening to them.

In depression, a primitive dread or hibernation reaction was evolutionarily adaptive. Mr. Caveperson may have experienced this reaction frequently given the dangerous nature of his environment but certainly as winter approached, which would be an inheritance from his animal (or even plant) ancestors. Although we've neutralized many of these dangers, the reaction lives on inside of us and may be triggered by less-dangerous situations or those with a symbolic value. Or it may simply be triggered by the waning intensity of the sun in fall. Those situations drag the unipolar (and bipolar) back to the body's reaction of neurotransmitter withdrawal, and thc individual takes on the typical symptoms of depression. His body is dangling in the river since dopamine is not well controlled, and lacking the gating function, he thinks in a primary process manner as did cave-persons although now he has words. Thus his concentration is poor and it takes great effort to pull himself out of the water, suppress dopamine, and think—the mode of mental function rewarded with those fertilizing miracle growers of the brain. He may sink even deeper into the stream and be prone to psychotic phenomena. One could define mood disorders as avoidant of secondary process (dopamine-suppressive) thinking in favor of primary process thinking.

Depressives' sleep, appetite, mood, energy, sense of self-worth, and enjoyment are all diminished by the waning of the neurotransmitters serotonin and norepinephrine. All of these qualities foster staying put in one's cave. The mind is preparing for a long winter with a need for minimal food, sex, and activity. Their minds are on autopilot as the waters rush around them. Dopamine is largely unsuppressed, leading to anxiety, insomnia, low appetite, and lack of enjoyment of anything as well as sensations of worthlessness and hopelessness. (The autoreceptor, that thermostat that turns on to reduce the flow of neurotransmitter, has shut down to try to encourage more neurotransmitter production in response to the deficit. Unfortunately it doesn't succeed.) Yet they don't let go of the branch completely, which distinguishes them from schizophrenics. What pulls them out of this morass are neurotransmitter promoters like serotonin reuptake inhibitors or norepinephrine reuptake inhibitors. They promote neurotransmitter function by blocking their reabsorption into the neurons that secreted them, increasing their concentration in the synapse channel between the nerves. This turns on the autoreceptor whose job it is to modulate the normal level of neurotransmitters, and it is this dance between autoreceptor awakening and neurotransmitter modulation that brings the mind back to normal functioning, which may take a few weeks to reestablish. ECT (electro-convulsive therapy) also helps, probably by resetting the dopaminergic stream and in effect suppressing it. (ECT probably also surges serotonin and norepinephrine, reigniting the dormant autoreceptor, which puts a brake on dopamine. And finally it strengthens the gating function markedly I suspect.) The raging river slows to a trickle, making it easier to pull oneself out. Alternatively it raises seizure threshold, possibly reducing the likelihood of triggering the primitive dread reaction. Mania can be prevented by anti-seizure medications that make it harder for the brain stem to go into rapid-fire mode. We use antipsychotic medication also in mania very successfully to dampen down the raging waters of dopamine.

There's another treatment that helps depression that I should mention at this point. It's called *sleep deprivation*. If the sufferer of depression stays up for a full thirty-six hours, they have a blistering rush of neurotransmitter and are much relieved. They cannot nap at all, and if they do, the effort is nullified. After the thirty-six hours, they can resume their normal sleep pattern. Some people maintain the benefit indefinitely, others relapse almost immediately. If it's repeated two or three times in a week, it tends to last. Sleep deprivation sets off an internal rush of neurotransmitters including serotonin, norepinephrine, and our gnarly friend dopamine. Let's combine this with some other facts and see if we can come up with a cogent theory, you and I, real scientists making hypotheses based on data.

Depression's typically worse in the morning. As the day goes on, things lighten up a bit. It also tends to be worse in winter, and in May the suicide rate spikes. Can we put these facts together and come up with a theory? We'll need one more thing to add to our data (well, maybe two).

Now it's time to recall autoreceptors, the tiny thermostats that are attached to the cells to monitor how much neurotransmitter is in the intercellular space. If there is too much, they send a signal to the cell body to stop releasing neurotransmitter. So in our description of the primitive hibernation reaction, when we describe a global pullback of neurotransmitter, we can assume they are involved. These guys are the perfect ones to orchestrate a neurotransmitter pullback. Suddenly you start to feel weak, your appetite is gone, you can't sleep, and you feel rotten. This signals the fall preparation for hibernation of Mr. Caveperson or, in some cases, us if we have major depression.

Antidepressants work by boosting the levels of serotonin and norepinephrine (maybe dopamine too) in the extra-cellular space of nerve cells in the brain, and they do that immediately. The different types of antidepressants all do this with different mechanisms of action. The original type of mood boosters, monoamineoxidase (MAO) inhibitors, were found serendipitously. Patients were being treated for tuberculosis. These patients, in rather dire circumstances and usually in quarantine, were noted to be fairly happy. They were laughing and playing cards in their confinement. It was found that some of the treatments for TB had MAO-inhibiting properties. MAO is a chemical in the brain that breaks apart neurotransmitters like serotonin and norepinephrine. If you inhibit the chemical that destroys neurotransmitters, you will have more neurotransmitter in the synapse of nerve cells. Voila! A new type of medication was born. (Serendipitous discovery, or accidental observation, is how many medications are born. There are many other neat examples of serendipitous findings in our pharmaceutical history.)

The problem is, if you eat fermented foods like cheese or yogurt that have lots of the chemical tyramine in them, a sudden, dangerous spike of blood pressure can occur as that chemical's destruction is blocked as well. There are two types of MAO, one of which is in the stomach. So in essence you were giving a person a means of killing themselves just by eating a few slices of pizza. The blood pressure spike is massive enough that it can lead to stroke, and MAO inhibitors have other side effects like dizziness. Further, as new medications arose, like Prozac, we found that we couldn't combine them with MAO inhibitors or various problems might arise including something called *serotonin syndrome* where you simply have too much serotonin. However, bottom line, they worked and worked well under the strict dietary circumstance in which they could be used. These have names like Nardil and Parnate, and selegiline, which is also used for Parkinsonism and comes in a patch that avoids the gastrointestinal system all together.

The next antidepressants to be discovered, the tricyclics, were named for their three-ring chemical structure and did not require that dietary precaution. They had plenty of other side effects like, oh, blurry vision, constipation, cardiac conduction abnormalities, urinary retention, dry mouth, and so forth. Yet they did and do work and work well not only for depression but for anxiety symptoms like panic attacks. How do they do the job? By blocking the reuptake of both norepinephrine and serotonin by the cells so that these chemicals accumulate in the synapse or intracellular space, boosting them. (You can see how all the antidepressants boost the neurotransmitters: they are unidirectional.) These are medications like Elavil, Pamelor, Imipramine, Sinequan, Vivactil, and Anafranil. They work great once you get used to them.

More modern antidepressants include selective serotonin reuptake inhibitors (SSRIs). Instead of blocking the reabsorption of norepinephrine and serotonin, they are specific to serotonin alone. In the early 1990s they were introduced and quickly took over the market. Why? There was no dry mouth, blurry vision, constipation, or cardiac conduction problems and, in overdose, SSRIs like Prozac were quite safe unlike the first two categories. If you're giving a depressed person medication you'd like it to be safe in overdose, wouldn't you? Also safe in overdose are antipsychotics. Valium-type drugs are somewhat safe in overdose *alone*, but when combined with other things like opiate pain killers or alcohol or other sedating medications, they can be quite deadly. We try to get people off benzodiazepines as quickly as possible. They are addictive and basically alcohol in a pill. They require withdrawal maneuvers in higher doses with a gradual taper. They are sold on the street and abused. Please don't resist your doctor if he says it's time to get off your Valium, Klonopin, Xanax, Ativan, and so forth. He's doing you a big favor. (This is not medical advice as each person's situation is different. You must discuss any change in treatment plan with your doctor who knows your particular case and history.)

Prozac quickly soared to the top of the antidepressant charts for its safety and general efficacy. Many of us believe the tricyclics are a little more effective, but if Prozac and its relatives like Zoloft, Celexa, Paxil, Luvox, and Lexapro work well, why not? Prozac is related chemically to Benadryl, an over-the-counter antihistamine, but Prozac doesn't make you tired or give you dry mouth. It can cause some jitteriness and affect appetite a bit. SSRIs have sexual side effects including decreased performance and decreased orgasm in up to 50 percent of patients. This is disturbing enough for many people not to want to take them. But they work relatively well and safely though at least one-third of those who take them (or any antidepressant) don't respond. They block the reuptake of serotonin alone leaving more in the synapse.

There are other antidepressants that I have yet to mention. For example Remeron and Wellbutrin, which don't fall into the usual categories. Remeron is an antidepressant that is very sedating, and we use it to promote sleep in depression but its mechanism of action is unclear. Wellbutrin is a stimulant type of antidepressant thought to work on dopamine and norepinephrine. It is safe but initially had an elevated seizure rate (all antidepressants can increase risk of seizures slightly) at higher doses, and the dose recommendation was lowered. Desyrel is a weak antidepressant primarily used as a sleep aid.

Norepinephrine reuptake inhibitors like Cymbalta have a reuptake effect on norepinephrine, not serotonin. Effexor is norepinephrine inducing at higher doses. It can also raise blood pressure at higher doses. These are some fun facts about antidepressants. But the principle is to find one that works for an individual patient and to adjust the dose to maximal effectiveness with as few side effects as possible.

Why does it take three to six weeks for these medications to reach their maximum effect? Very smart question. The usual explanation has to do with autoreceptors. The theory has to do with the assumption that when the flood of neurotransmitters is in the synapse, which happens as soon as you take the antidepressant, an autoreceptor that has basically shut down in the face of neurotransmitter deficit, true to form, turns back on, sending a strong signal to the cells to slow the production of neurotransmitters. The cell body listens, so the flood of neurotransmitters goes back to a premedicated level due to autoreceptor inhibition. But then somewhere around three to six weeks after the antidepressant is started, the autoreceptor tones down a bit and gets into a dynamic balance with the level of neurotransmission. An attractive theory. Whether or not it is true is uncertain. Most psychiatrists tend to believe it. The bottom line is that the job of an antidepressant is to turn the autoreceptor back on. If the level of neurotransmitter is too low, it can't. And turning it back on helps us suppress and regulate dopamine. This allows for the normal modus operandi of the brain—ego stimulation of the forebrain with conceptual material—to resume and secrete BDNF and VEGF, the brain expanders. If dopamine is resurgent, as when the autoreceptor is shut down, concentration is impeded as the gating function of dopamine suppression is reduced and we cannot think. We need solid gating to think clearly. If there is rupture of gating the mind blocks the think space with manufactured verbiage.

So how do we put together a theory that explains why (1) depression is worse in the morning, (2) depression is usually worse in winter, (3) May is the highest suicide month, and (4) sleep deprivation gives one a strong euphoric boost? This is where the what-ifs come in. Science progresses by these what-ifs. They are not necessarily expected to be correct. Science is full of questions and sometimes short on answers. Hypotheses and theorizing are how we advance even if lots of theories are ultimately discarded. Does this remind you of anything? Evolution! Evolution advances by mutational hypotheses, discarding most but keeping the ones that further our genetic knowledge.[1]

Well, here goes, let's try this. What if the autoreceptors burn out each day as the day goes on and then reset at night and are full strength in the morning? Hmmm, uh, what would that mean? Well, our first fact that depression is classically worse in the morning fits well. If autoreceptors reset at night and peter out during the day, they would be strongest in the morning. Don't forget that their job is to send a tone-down signal to the cell to stop producing neurotransmitter. Reduced neurotransmitter means more depression. And we treat depression with chemicals that invariably boost the level of neurotransmitter in the space between the cells.

So far so good. Next fact: depression tends to be worse in winter. Winter with its reduced light is a time when Mr. Caveperson either hibernated or felt like hibernating. When the sun went down he had no artificial light so he went to sleep. Assuming lets say that the autoreceptors weaken in the afternoon, lifting our spirits, in winter we spend less afternoon time awake and uplifted and instead more time asleep. But you say quite keenly, modern humans have artificial light and can stay up late into the wee hours even in winter. Don't forget that artificial light is not sunlight, which we call *full spectrum*. People treat so-called winter blues with full spectrum light by staring at a light box in the morning. This helps, but our regular artificial lighting does not. It's as if we go to sleep once it's dark, and the artificial light we stay up with is not effective. Plus our ancestors for millions of years met winter with more sleep, a hibernation state. This pattern is no doubt ingrained in our ancient mental wiring. The first two facts are nicely explained by our theory.

Next, May is the highest suicide month. Aha! you say, our theory now falls apart. After all, as the light increases in the early spring months, affectively disordered people (which means uni- and bipolars) should start to feel effortlessly uplifted. They should sleep less and experience the joy of afternoon autoreceptor fizzle. Sounds like our theory has taken a hit. Here's the old explanation as to why affective disorder patients tend to kill themselves in May. The concept is that if the melancholy are depressed all winter, they lack the energy to carry out their suicidal plan until the spring when the increasing light gives them a boost of energy and they get up and kill themselves. Hmmm. What do you think of it? Well I've treated thousands of affective disorder patients. Those winter depressives, to a person, were not so depressively drained of energy that they couldn't get up and find a means to try to kill themselves.[2] So that theory, at least in my book, doesn't hold much water.

My theory goes something like this. Each fall and winter, uni- and bipolars fall into a rather steady depressive gloom. They may not like it but at least it is predictable and lends itself to a denial that they are ill. As spring starts to set in, they notice an affective episode with various symptoms such as poor concentration, sleep difficulties, change in energy, difficulty enjoying things. In short they experience a dysregulation. The change in mood isn't a straight road upward; it consists of a craggy climb. This reminds them that they have an illness and breaks through their winter denial. At this point they start to despair and become somewhat hopeless, which often precedes suicidality. It's basically saying, "What's the use?" Spring is then a reminder that one has a significant brain disorder that puts us out of control of our mental state.

Well all right, our theory stands although some skeptics might disagree with my last explanation. Blessed are the skeptics, but I think it makes a lot of sense. Our final bit of information, that sleep deprivation leads to a quick mood boost, fits perfectly with our theory. The depressive can remain awake for many hours (assuming they manage to avoid sleep—no mean feat) when their autoreceptors have reached the state of burnout, meaning that they are in an undampened neurotransmitter state for hours longer. It truly does work, according to the research. I have often wanted to try it but each time I get close to doing so I chicken out nor have any of my patients been so inclined. There is plenty of research on it however.

So I think we've found a theory that explains the facts as we know them. If autoreceptors burn out as the day goes on, we experience a mood lift toward afternoon, and then when we sleep they reset and are in full force in the morning, drowning us in depression in the a.m. hours. Let's pat ourselves on the back and say that together we have advanced science a little bit. Unfortunately we will have to leave it to others to research this theory if they so choose.

So to recap, we can equate depression with the hibernation or primitive dread reaction that Mr. Caveperson experienced in winter's fading light and in the challenging landscape in which he or she lived. It was a daily struggle to survive with endless risk, skull-crushing brutality, and few rewards. One might ask if primitive people were happy at all and if their usual mood was something akin to our major depression. This is probably an undiscoverable notion as moods do not leave fossils. Yet we know that the adoption of language gave our minds an entirely new context. The average state of happiness of modern beings was not possible millions of years ago. Civilization has neutralized many of the ubiquitous predators that our forebears did combat with routinely. One could say that happiness is a modern invention, an evolutionary masterpiece that erupted with the expectation of survival and the cultivation of leisure. Thinking, which arose from language in the sapiens only, releases BDNF and VEGF, which are euphoriants as well as brain fertilizers. The superego, a recent outgrowth of civilization's social necessities, contributes to our state of well-being, which was not the case in pre-civilized lifestyles. We reward ourselves for following the golden rule. These modern developments did not effect the Neanderthals' state of mind.

So theoretically, when one is in a major depression one is returning to a state of mind hardwired into the primitive brain either to induce bear-like hibernation or respond to something dire that's happened. In us moderns, this is accompanied by dopamine de-suppression as our friend the autoreceptor shuts down, embracing the usual underlying theme of increasing entropy and eschewing secondary process thinking in favor of primary as the gold standard of complexity (contemplative thought) is dissolved in a sea of faltering concentration and raging dopamine. The brain responds by vacuuming neurotransmitters, a signal of the gravity of the situation, and our modern antidepressants, flawed as they may be, can reverse that pullback by streaming lots of neurotransmitter into the synapse. This turns the autoreceptor back on as any treatment for depression must, restoring the contemplative framework. Cognitive maturity and sexual maturity also play roles in happiness. The joy children experience, while delightful, is more superficial and transient than the happiness of adults. If Mr. Caveperson had joy it was in vanquishing a predator, experiencing the joy of reproduction, or relieving himself of hunger or pain.

But there's another treatment for depression that is very safe and effective: ECT. We should review a bit about ECT, a much maligned treatment. To go back a little, someone came to the conclusion a long time ago that epileptics don't get depressed. *Wrong*. But some of our greatest treatments were probably built on false premises or serendipitous findings. But, desperate for depression treatments (antidepressants weren't available until the 1960s), scientists in the 1930s experimented with different ways to induce seizures. They tried chemicals like Metrazol, and they tried insulin to induce hypoglycemic seizures. These weren't reliable. Along came two guys from Italy, Cerletti and Bini, who pioneered the use of electricity using small electrodes on either side of the head to induce seizures. Lo and behold, the effect was astounding. People who had been depressed for years, including schizophrenics and uni- and bipolars, got up and started living again.

But the treatment suffered from its effectiveness and lack of alternatives and was probably overused. Tons of people in large state hospitals regardless of diagnosis got ECT because there was only one other treatment: lobotomy. If someone offered you a choice between having electrodes placed on your head and then being given a shock to cause a seizure or having your forebrain liquified by a surgical needle, most of us would choose the former. Lobotomy may not have been a proud moment in psychiatric history, but bear in mind it was an attempt, albeit desperate, to help suffering people when little else was available. Probably it quieted the sinister agitation of schizophrenics by ending the conflict between the primitive organization and the modern brain in favor of the former. In the early evolution of the illness, the battle between the twenty-year-old brain of the schizophrenic and the prehistoric primitive organization scorches the mind with positive symptoms. The resolution may take years of torturous combat but it could be suddenly resolved with a pointed lobotomy probe.

So, yes, ECT was overused. But it helped an awful lot of people without causing brain damage. It was also more barbaric back in the early days. No anesthesia was used and no muscle-paralyzing agent. The patient had to be held down. The seizure caused intense contraction of the muscles and therefore spinal fractures occasionally ensued. The electric voltage delivered was blockbuster crude.

The procedure now is much more refined. First of all the patient is asleep under brief anesthesia. Before the electric shock and only after being fully anesthetized, they are given a muscle-paralyzing agent preventing the intense contractions of yesteryear. Using modern techniques of square wave and brief and ultra-brief pulse electric charge design, the total current used is much less while resulting in a reliable seizure. The general anesthesia probably poses more risk than the procedure itself. However, this is one of the safest procedures in all of medicine. Jaw aches from biting down (we use a bite block to soften it) and headaches are two common side effects. Yes there is memory difficulty for some but it is transient. Many very highly functioning people have had ECT. It prevents suffering. It's not for everyone, but if your psychiatrist says you need it, don't just dismiss it out of hand. Consider it objectively.

It's usually done three times per week up to about twenty treatments, and you start to feel better after the first three. If your memory gets bad the electrodes can be switched from bilateral (on either side of the head) to unilateral (one on the forehead and one on the right temple). This is done because the right side is typically the nondominant side since most of us are right-handed, which indicates that the left side of the brain is predominant.[3]

"So how does ECT work?" you ask. Well, we don't exactly know. That's reassuring, isn't it? There are many theories. ECT has also been used for tardive dyskinesia, intractable seizures, and Parkinson's disease. Here are some thoughts.

When you have a grand mal seizure, there is a global depolarization of the brain. When nerve cells fire they depolarize, meaning certain ions like calcium rush into the nerve cell while others like potassium and sodium are pumped out, decreasing excitation, increasing entropy. This is how neurons fire, by suddenly opening their skin so to speak to let ions in the blood rush in or out. That's one reason not to be dehydrated or low on potassium or other ions that could affect the firing of such neurons including those that conduct a signal to the various chambers of the heart. Orderly conduction of nerve signals is important. The electrodes placed on either side of the head in ECT produce enough of a disturbance to depolarize the brain globally.

Okay so what does a global depolarization of the brain do? It's like shutting your computer down and then turning it back on and letting it boot up again. Everything goes back to factory settings. So for example, if you have an excess of dopaminergic flow, that will normalize. If you suffer from a deficit of neurotransmitter activity, that will normalize, possibly by returning the autoreceptors to usual functioning. Don't forget, in depression the autoreceptors have turned off yet there is a deficit of neurotransmitter leading to a feeling of dread and sadness. And given that in both depression and mania dopamine will be theoretically excessive, ECT would normalize those dopamine levels. We use antipsychotics to treat mania and at times depression. These, as you all know, block dopamine receptors. Ultimately ECT strengthens gating, the final common pathway for poly-diagnostic improvement.

Over the course of ECT, the individual's seizure threshold goes up. With each treatment the threshold required to push their brain into a seizure increases a little and we often have to turn up the energy as we proceed. What do we do if we get to the highest setting the machine will allow (set by the FDA) and we don't get an adequate seizure that is at least twenty seconds long? We add intravenous caffeine! What? That morning coffee you drink lowers the seizure threshold significantly enough to increase the length of the seizure. The average cup of coffee has about 120 milligrams of caffeine. We can use up to, oh, 1,000 milligrams or more of caffeine to lower the seizure threshold and help the electricity induce a longer treatment. Why does seizure threshold go up in the first place? Perhaps with each seizure the brain is learning to reset itself in a way that attempts to protect it from what just happened, a grand mal seizure. (In case you're wondering, yes, there are petit mal seizures and other types. Petit mal usually consists of staring spells, blinking, and feeling out of it, but this can vary.) What else raises the seizure threshold? Anti-seizure medications like Depakote, Tegretol, Lamictal, and so forth. We use some of these medications to treat bipolar illness quite effectively by blocking mania, which may rely on lowering the seizure threshold to boost general level of excitation in the basic parts of the brain like the brain stem. ECT is also effective for bipolar illness.

So why, if ECT causes some memory difficulty during the course of the treatment, does moving the electrodes to the nondominant hemisphere reduce that risk? Well, language is located in the dominant lobe of the brain, typically the left. If we shift the electrodes to the nondominant hemisphere, we bypass somewhat the language center at least with the main thrust of electricity. Let us assume that ECT makes it a little harder for the brain to connect words to their targets. This happens, as previously mentioned, in the preconscious. This is probably why there are word-finding difficulties during the course of treatment that then clear up once the treatment is finished. Shifting the electrical charge away from the dominant language center may minimize that. And in lowering overall excitation of the brain we have successfully gratified our old friend entropy's desire while re-suppressing dopamine, restoring the ability to think. (I am not recommending anyone have ECT. That is between you and your doctor.)

Now let's take a quick look now at obsessive-compulsive disorder (OCD). It's a disorder that results in compulsive behavior like repetitive checking of door locks or that the water is turned off or that there are no lights on as you leave the house. Obsessions are repetitive, unwanted thoughts. It so happens that OCD has a high incidence of Tourette's syndrome. This is a tic disorder in which the person has involuntary movements, like sudden jerking of the neck or stomach muscle twitching. Some may call out loudly at times or even curse. These are generally involuntary movements but Tourette's sufferers will tell you that they can mold or delay them with effort. OCD patients can also modify their compulsions somewhat but if they are kept forcibly from doing them they get very restless and anxious. What do we make of this comparison?

Well, tics arise from an imbalance of dopamine and acetylcholine. If dopamine is too strong, the acetylcholine has to break through the dopamine blanket with a tic-like punch, which is why we use dopamine blockers to treat it. This jerking movement is involuntary and imperative but can be delayed with conscious effort. Let's assume that in OCD there is an imbalance of the dancing pair dopamine and acetylcholine as well. Instead of an involuntary gratification of an acetylcholine urge, there is a voluntary gratification of it in the repetitive compulsive behavior. Don't forget, movement suppresses dopamine. In order to accomplish this, there has to be a thinking component along with it to persuade the patient to perform the act, to grease the wheels so to speak. For example, "If I don't check and recheck the front door I'll be robbed of all my possessions" is a (delusional) thought that might encourage a compulsion to check the door locks multiple times, which has an urge from acetylcholine behind it but not as focused in the motor (nigrostriatal) tract of the brain. Thus in moving away from an involuntary movement, a cognitive (false) belief is enlisted to motivate the sufferer to gratify the acetylcholinergic urge to burst through an excessive dopaminergic tone. Thus we have altered the situation only slightly. To carry out the voluntary compulsion, an obsession is used to motivate the obsessive compulsive, which then gratifies the acetylcholine impulse. This is one step away from an involuntary tic but with a shift to a different brain tract that is less involved in motor impulses and more engaged in thought production, which of course requires language.

Where else do we see this kind of effort? Schizophrenia. In this disease, delusional thinking alters the patient's behavior in favor of the primitive organization. For example, "I don't have to pursue my career as a research biologist as I am the second coming of Christ and will feed the starving children of Africa." Those who try to dissuade the schizophrenic may be seen as evil and thus avoided. In both cases, the false belief serves the purpose of the illness taking over voluntary musculature and furthers the suppression of dopamine. One might say that control of voluntary muscular is a prize goal of schizophrenia and part of an inexorable effort on the part of the primitive organization to control the actions of the patient. It usually succeeds. So the symptoms of OCD while reinforcing dopamine suppression may be a sacrificial lamb offered up in an attempt to prevent a full-blown schizophrenic intrusion. Don't forget that dopamine suppression originally started in the brain system involved with movement, the nigrostriatal tract. But talking itself is a motor function and one that is intimately connected with ideas, thinking, expression of concepts. Thus the dopamine suppression shifts to the central part of the brain as a general trait as well. In schizophrenia the suppression of dopamine is reversed. In an effort to stop this reversal, people will adopt certain symptoms as in OCD that encourage dopamine suppression enhancing acetylcholine. Other symptoms like self-cutting may also be attempts to avoid complete schizophrenic dopamine resurgence. Major depression with psychotic thinking also may serve the same overall function, the principle being "Give the schizophrenia part of what it wants and maybe it will leave me alone." Remember our friend Schreber engaged in such a bargain by adopting the belief that he'd been turned into a woman in order, he assumed, to avoid the mental deterioration of schizophrenia's scourge.

We treat OCD with serotonin-type antidepressants in high doses. By doing so we turn the autoreceptor back on. This has a dopaminergic suppressive effect that then allows the dopamine-acetylcholine balance to restabilize more equitably. We can also use low doses of atypical antipsychotics to help suppress dopamine. In short all antidepressant treatments *turn the autoreceptor back on*. This allows it to tone down dopamine (and dopamine can lead to the demons of anxiety, insomnia, poor appetite and concentration), which in turn allows for the normal functioning of the executive forebrain with secondary process type thinking and the release of BDNF and VEGF, the brain fertilizers that cause nerve cells to flourish and expand. This is the normal state of operation of the healthy brain. All depression treatments must do this to succeed. Ketamine stimulates BDNF and VEGF in the forebrain, inducing euphoria. So the major takeaway here is that OCD is only a step removed from a motor problem like Tourette's. It is intimately wrapped up in the acetylcholine-dopamine dance. If schizophrenia is the ultimate in dopamine de-suppression, some mental infirmities use the trick of promoting dopamine suppression in hopes of avoiding the major mental disorder that is schizophrenia. Obsessive compulsives spend a lot of their time carrying out these rituals, which are motor attempts to dart out of schizophrenia's grasp as it encourages acetylcholine over dopamine.

Occasionally art has an effect on a medical treatment or on political movements or fashion or culture significant enough to damage or promote or simply change it. So what am I talking about? A movie called *One Flew Over the Cuckoo's Nest*, which was originally a book by Ken Kesey, had a significant effect on ECT. This movie was released in the seventies. The sixties and seventies were an age of rebellion against authority (e.g., the Vietnam War). Individual human rights were on a pedestal. ECT had been in use since the thirties and maybe, as I've said, it was overused. It was also barbaric without anesthesia and muscle paralysis so patients had to be held down to receive the initial shock. Well, the setting was ripe for a film like *One Flew Over the Cuckoo's Nest* to come out and portray ECT as a mind-numbing, vegetating, abusive technique. Starring none other than Hollywood's bad-boy super-talent Jack Nicholson, the movie portrayed ECT as nothing more than authority's way of punishing and destroying the minds of inmates.

Bottom line, ECT fell into a decades-long hibernation. It was banned in some counties of California. People who clearly needed it refused it and missed out on an opportunity to get better faster. This has slowly changed, however, and since the eighties or so ECT is back to what one would call *rational use*. Most states now have a legal procedure to force psychiatric patients to get ECT if their physician can prove to a court that it is the only thing that will save the patient. Examples of such a situation include patients who are so depressed they've stopped eating or drinking, patients so intent on killing themselves that even a locked psychiatric unit cannot stop them, and manics so out of control that they're likely to harm themselves or others. In cases like these, the judge has the right to order ECT. But the point is, art knocked ECT out of rational utilization for decades. That's dangerous. There's nothing wrong with skepticism, but one must try as much as possible to keep an open mind.

Imagine you were allergic to a state of mind. Every time you got antsy or inquisitive or lonely you broke out in a rash, your heart raced, and you had difficulty breathing. Well, manics are in a sense allergic to depression. Most bipolars initially have several depressions. They find this state of mind extremely painful and, what's more, a decided reminder that they are not in control. Loss of enjoyment, loss of motivation, and loss of self-esteem are devastating. Somewhere along the depression trail their minds learn to cope with it by overriding it. Sluggishness becomes super energy, loss of esteem is now grandiosity, and poor sleep becomes no sleep. Wild impulsivity and thrill-seeking replace a staid, withdrawn sentiment of depressive hibernation.

The manic is in a frenzy to avoid all that depression is. It's as if it has taken several depressive detours for them to learn the art of manic avoidance. Eventually it becomes automatic for them to launch into mania as soon as a depressive shadow falls over them. If we view depression as a primitive hibernation state of mind, how should we view mania? Instead of a life-threatening event triggering a gnawing dread, manics view depression itself as a life-threatening event. There is a desperate attempt to override all that depression is. Instead of sinking into dreaded hibernation, they launch into a fight-or-flight explosion with adrenal gland stimulation and what may turn eventually into a dopaminergic psychotic hailstorm. They may be extremely impulsive, agitated, or paranoid and do very self-destructive things. It's as if they take bad news as a challenge. "I am not going to succumb to this feeling of dread," the manic says, consciously or unconsciously to themselves, and this becomes an automatic response to its onset. However, once the mania collapses, which it invariably does as it is not sustainable, the depression ensues. Bipolar patients have a high suicide rate as their schema to avoid the morass of depressive insults fails.

Here are some facts about bipolar disorder (manic depression). It usually starts in your mid-twenties to thirties—later than schizophrenia. Episodes of mania may last days to months. What we call *bipolar 2 episodes* are shorter and less typical in symptomatology. They may simply involve an energized period of beefed-up irritability with some insomnia and mind racing. A full-blown, bipolar, type-one manic episode often involves racing thoughts,[4] lack of the need for sleep, and euphoric ebullience (manics enjoy this state of mind and will often resist treatment for this reason).

Once the treatment, which involves blocking dopamine, and raising the seizure threshold succeeds in taming the mania, depression may ensue. When manic, these patients talk, talk, talk and their nonstop recitations can be quite difficult to redirect. Impulsivity can be problematic and may involve sexual indiscretions, poor financial decision making, or minor illegalities like shoplifting, gambling, alcohol, or drug use. This can get the manic into a lot of trouble: for example, when they max out their credit cards and borrow from their kids' college fund to finance some half-baked scheme to start a business that is destined to fail. Manic grandiosity makes them feel like they're functioning super well but in reality they are not. They're extremely labile, one minute angry, the next minute sad, the next happy. It's a Coney Island of the mind. (*A Coney Island of the Mind* was a book by Ferlinghetti. I don't know if he was bipolar but many artists are.)

Treatments for mania other than dopamine-blocking agents often involve mood-stabilizing medications including lithium, Depakote, Tegretol, Lamictol, and Trileptal. Of the ones listed, only lithium is not an anti-seizure medication. The rest were all originally used to reduce or prevent seizures. This means they raise the seizure threshold as does ECT. Does this imply that bipolar mood swings from deep depression to euphoric mania are seizures? No, but one could argue that they have a similar basis. It's likely that mania rests on lowering the depolarization of brain cells in the brain stem, which controls basic functioning like respiration and heart rate, making it easier to plunge into a revved-up universe of manic excitation. Anti-seizure medications would increase repolarization, making it harder for one to reach the manic state of euphoric infusion. Lithium, an oddball mood stabilizer, is a puzzle that no one has figured out yet. How it works exactly we don't know but it seems to have a nerve cell stabilizing effect. At one time European doctors would treat bipolar illness by recommending patients take warm mineral baths in specific vacation spots. These baths were found to have a high lithium content. Lithium was used in Europe before it became available here. It had an unfortunate stint as a salt substitute, which caused deaths because lithium needs to be monitored closely and not just sprinkled wantonly on soups and fries. If used properly it is one of the best mood-stabilizing agents and it also has a specific anti-suicide effect. When patients go off lithium, their suicide rates go up. Lithium is a basic element and is in trace levels in the water supply. In communities where there is an abnormally high level of lithium, suicide rates are lower. Could it be then that lithium levels were higher in the evolutionary primordial environment? Possibly. If we evolved in a higher-lithium-level world with levels subsequently declining as rain and sea boundaries changed, today we may all be somewhat lithium deficient. Those of us with mental illness could be more susceptible to this deficiency. But of course no medication is for everyone. Some manics eschew lithium possibly because it is just so darn effective in preventing their mania. As I said, it must be monitored carefully. High blood levels of lithium can cause diarrhea, then wooziness, then coma, and then death. It may have a slight effect on the kidneys so renal function must be checked yearly. It can certainly affect thyroid function, which must be checked periodically.

What makes treating bipolar patients so difficult (besides patient resistance . . . Who wouldn't want to be in a state of euphoric grandiosity?) is the balancing act that goes on between antidepression and manic control. Antidepressants with their neurotransmitter-enhancing properties tend to induce mania. So to treat a depressed bipolar patient we want to introduce antidepressants cautiously and at lower doses. And we want to use them in conjunction with a mood stabilizer or an antipsychotic that puts a lid on the mood expansion. It may take some trial and error to find the right balance.

The manic dances between two lofty poles, one rampant and uncontrolled, the other leaden and ashen with immobility. Neither suffices, however, and the potions that drive a freeway down the middle permeate the loss of one and the avoidance of its opposite. That middle course is the desired option—not necessarily the most enticing but stable in its rectitude and civility, navigating its riders between Scylla and Charybdis. The ploddingly sane, eschewed by discontented manics everywhere, is the lofty goal of our witches' brew of cures, making them somewhat unpopular.

Getting back to our branch-and-stream analogy, bipolar patients are hanging on to the branch but their bodies are dabbling in roaring, dopaminergic waters. Sometimes they lose their grip on the reality tree and get swept away into full-blown psychosis. What contributes to this raging river? Sleep deprivation. Remember we discussed using sleep deprivation to treat depression? When manics stop sleeping, their autoreceptors remain shut down and dopamine is released intensively, resulting in a psychosis akin to that of a methamphetamine user. They can become quite paranoid and agitated since they're talking a mile a minute as they avoid secondary process thinking, and they may start to view sleep as the enemy. It's just a matter of time before some incident occurs making it mandatory for them to get treatment, frequently in a hospital. Often their family will insist they get hospitalized to prevent an encounter with the police. It's not surprising that antipsychotics, which block dopamine receptors, are effective in damping down mania. The raging stream slows to a trickle, and the manic can pull themselves back on to the branch and regain a sense of reality. Once they start sleeping, the neurotransmitter rush subsides as autoreceptors are turned on during sleep and there is less dopaminergic pressure. Thinking can shift from a racing, primary process, dopamine surge that pleases entropy to a slower, secondary process deliberation as the autoreceptors dampen down the dopaminergic flood with the help of this chemical blockade. It is not surprising that all diagnoses of mental illness have increased suicide rates . . . Their proximity to entropy lures them to the ultimate entropic victory, returning to inorganic matter.

So how, one might ask, if the manic is responding to dopamine blockers, if they're psychotic and agitated and revved up, is this different from schizophrenia? Would it surprise you to hear that for decades there was no clear distinction between schizophrenia and bipolar disorder? Psychiatrists debated this issue hotly. What was referred to as manic depressive insanity suggested a dividing line between that and schizophrenia. We still to this day have trouble at times making the distinction. Esme Wang, author of *The Collected Schizophrenias*, was initially diagnosed with post-traumatic stress disorder (PTSD), then bipolar disorder, and then schizoaffective disorder. The gradation can be very challenging. But we are now in a position to clarify this age-old confusion. Using our branch-and-stream analogy, the schizophrenic gets swept away by the raging dopaminergic water and remains in them. Even when the waters subside, he or she may not get back to the branch and pull themselves up since the dopaminergic river has carried them too far downstream, their gating function fails, and the primitive organization reestablishes its hold on the mind, analogous to fission where a massive amount of energy is released gratifying entropy's desire. Bipolars, on the other hand, get dunked in a raging stream of dopamine but can pull themselves out again once a dry spell hits and the gurgling rapids recede. Their functioning between episodes can be fairly normal—unlike that of most schizophrenics. They have not suffered the cognitive regression seen in schizophrenia. Between episodes, bipolar patients use the same cognitive rules as any adult. So if we had to make a dividing line between bipolar disorder and schizophrenia, it would be the battlefield of cognitive dysfunction. Schizophrenics are mired in cognitive regression akin to childlike expression and thinking 1.0 or the thinking of intoxicants. Bipolars are not, at least between episodes, and many bipolar patients are highly functional. There are many creative people who are bipolar, and the manic high that possesses them at times can be used as a super-functioning state. They can get a lot done in a very creative time period full of overconfidence although what they do is often of lower quality. This is usually followed by deep depressive submersion. These cycles can be quite debilitating.

Stepping back for a moment, we can try to organize the affective or mood illnesses (uni- and bipolar) in relation to evolution. If we bear in mind the changes to the human brain wrought by language, depression may be regarded as regression to a more primitive and entropic emotional state that avoids secondary process thinking. The pre-verbal human had a less-organized, more-childlike mental habitat and inherited from our animal ancestors a tendency to hibernate in winter months. Reactions of euphoric victory over predators may have also been present, laying the groundwork for an effervescent mania. Once Homo sapiens advanced, they achieved a modern, adult-level of happiness coinciding with their enhanced safety and security—a product of civilization. Yet some are prone to regress to the old state of mind of our ancestors with dopamine de-suppression and a victory for entropy. While we less often need the repetitive dread reaction or hibernation with its neurotransmitter withdrawal, they remain there, hardwired into our system like an appendix. If schizophrenia is a return of the primitive organization and its cognitive regression, then uni- and bipolar disorders are a return to a primitive emotional state or, in the case of mania, an attempt to override it with a template in the euphoria of vanquishing a dreaded predator, which would also be evolutionarily favored. Either way there is a return to the primitive, a failure of dopamine suppression, a retreat from secondary process thinking, and the usual expected victory for entropy.

The failure of dopamine suppression can go down many pathways of expression. Depression is commonly linked with anxiety. Concentration is impaired as well. These are direct indicators of dopamine's surging influence and they improve once the antidepressant succeeds in turning the autoreceptor (which dampens down dopamine) back on. The dopaminergic flow can also express itself as pure anxiety without much depression, OCD, and ultimately mania. They all have a common root. In a different dopaminergic tract, the nigrostriatal, dopamine excess can be seen as Tourette's syndrome, stuttering, and Huntington's. Alzheimer's is probably an acetylcholinergic deficit; ADD, an acetylcholinergic excess. Dopaminergic deficit can be expressed as restless legs or of course as Parkinson's disease. We see that in renegotiating our relationship with dopamine we have benefited greatly and incidentally created a lot of dopaminergic collateral damage. We can control this indirectly by turning on autoreceptors with antidepressants, blocking dopamine with antipsychotics, stimulating dopamine or acetylcholine with other chemicals, and using treatments like ECT to reset the chemical balance of the mind to factory settings. The changes in neurobiological functioning wrought by language have not been entirely positive, but on balance, the benefits have far outweighed the drawbacks.

But this dopamine backflow or de-suppressive force allows us the pleasure of considering some enticing possibilities. Could it be, perhaps, that the schizophrenic inherits a general propensity to de-suppress dopamine? One might, then, encounter a panoply of global genetic stumbling blocks that lead to tributaries of potential diagnoses and not just one individual label. Indeed, we see families of schizophrenic patients with a variety of diagnoses connected with them. And we see that so-called polygenic risk scores are indicative of multiple diagnoses and not just the schizophrenic diathesis, leading one to consider a universal causality.

But it might also be worth a moment's consideration of the unusual scenario that the diagnoses we see in our patients are merely stopgap attempts to avoid schizophrenia, that the mind's counterpunch to a sensation of takeover is to elaborate one of the lesser illnesses as a sacrificial lamb to shift dopamine de-suppression away from schizophrenia. This hypothesis posits anxiety disorder, major depression bipolar disorder, OCD, and more as attempts to calm the raging bull of schizophrenic regression by throwing it a bone. Just as Schreber chose femininity instead of dementia (although he clearly had schizophrenia), mental illnesses may be compromise strategies designed to head off the flood.

* * *

Can one say that something as innate as happiness could be an invention? According to Harari (2015), happiness is purely a matter of neurotransmitters. Your genetically set concentration of dopamine and serotonin determine your level of happiness and that's that. Despite hardships and circumstance, one always returns to that happiness score.

Well, yes and no.

Major depression is marked by a neurotransmitter deficit. The medications we use all boost the level of these chemicals in the synapses of the brain to reverse the process. But my theory is that this neurotransmitter deficit stems from a prehistoric hibernation or dread reaction to lurking danger that is evolutionarily codified. If humans had no internal mechanism for feeling in danger, they were likely to go about their business of gratifying an urge for hunger, sex, shelter, and so forth, while ignoring potential disaster outside their door. With a bear pacing around your cave waiting for you, a sense of dread must ensue. Your need for food, reproduction, and activity should shut down in a hibernation scenario that might just keep you gloriously alive. Low energy, poor sleep, and a feeling of terror all contribute to the awareness of a grave potential nearby. That six-million-year-old mechanism might kick into gear in our modern era when it is no longer applicable. It feeds on itself as that depression magnifies small threats into Godzillas. But again, it is the return of the primitive dread or hibernation reaction that brings on the mental illness of depression in modern humans. There is loss of the modern installment module of happiness or what we call *anhedonia*, an inability to enjoy, and an increase in entropy accompanied the usual dopamine de-suppression.

Which brings up the question, did our fellow caveperson experience happiness? Did our precursor Neanderthal or Denisovan or Naledi experience the type of mood states that we moderns do? And what about suicide? Was that a regular outcome of the prehistoric diathesis? On the latter point, I would argue no. Prehistoric humans were too busy clawing their way to survival to consider suicide any sort of necessary option. Perhaps in a situation of imminent death there might be a decision to end one's own life one's own way instead of, say, by being ripped limb from limb by a surly gorilla. But apart from that, no, suicide was not a feature of the prehistoric human's repertoire. In fact, I would further assert that suicide can only be a facet of a modern society that expects happiness. And on that and many other bases, I suggest that happiness is a modern invention.

Perhaps in a situation of predator vanquish there was a giddy Neanderthal rush of victorious neurotransmitter. When they finally succeeded in spearing a lion to death and looked forward to eating it, there was a surge of euphoria. Similarly perhaps, after finding and consummating a mate, a Neanderthal might have a sense of joy created by an internal dopaminergic rush. But it was only in response to the successful achievement of evolutionary mandates that it occurred. And of course, in our dictum, this ancient celebratory revery turned into a template for modern manic euphoria. But with our newly minted language mechanism leading to thinking and contemplation with its deductive reasoning, modern humans neutralized most of those nasty forces of doom. We now expect to survive and take survival for granted. We don't have to leave the cave and slaughter a boar for the night's dinner. We are able to search for a mate in a more refined way that doesn't often lead to mortal combat with a dominant tribal figure. We expect a reasonable degree of contentment during our lives with inevitable ups and downs.

So again, my assertion is that Neanderthals didn't have true happiness. Well what about children or schizophrenics or the intoxicated? After all, intoxicated people often are in a euphoric high. Children laugh giddily at a clown on TV, and schizophrenics are known to be happy. So on what basis could I say that only modern-day adults experience true happiness? (I know I didn't exactly say that until now, but that's where this is going at the moment.) It takes a complex mental structure to truly feel the depth of joy that adults can feel. While a child may feel giddy, it's a superficial type of joy that is quickly replaced by the next emotion, often tears, boredom, or hunger. An adult human may encounter multiple layers of contentment based on factors that are quite complex and not just a brief, clown-like exhilaration. An intoxicated human, perhaps on a euphoric high, isn't experiencing what most of us would call true happiness. In fact there's a desperation in the intoxicant's behavior that reflects an inability to maintain any degree of true happiness. Manics are known to have a labile affect, joyous one minute, enraged or tearful the next. So in essence, what I'm saying is that true happiness arises within the context of a complex brain structure with all of its hierarchic organization, differentiation of parts, and prioritization of tasks and stimuli under the leadership of contemplation. The brain of a Homo sapiens drew in language like a sponge, and the language module changed our brains forever. We were now capable of contemplation and problem solving. Our minds complexified so that it became an automatic process of wordification of our inner mind process to reach an ability for universal communication. We were better able to differentiate ourselves from others thanks to something called ego, which strengthened Captain Kirk. The structural elements of our minds were enriched as we suppressed dopamine, gated out the primitive, strengthened our egos, and distanced ourselves from the entropic primitivity of our prehistoric ancestors. We became capable of much more specificity and detail. No longer was every four-legged thing that moved a cat. No longer was a hard, motionless object something to sit on, it was a huge, gray rock with moss all over it. We learned to suppress dopamine, and this talent led us to the modern ego landscape and the demise of primitivity although it still resides in us all and we revisit that entropic rabbit hole in sleep and intoxication. One of intoxication's greatest seductions is the return to a state of entropy otherwise unavailable to our minds while awake. The schizophrenic, however, in returning to the primitive organization, is similarly delivered to this anergic level.

So it is clear, at least to me, that happiness truly arises within the context of a modern, complex brain structure rather than one of a child, a prehistoric person, or an intoxicated soul, and that the type of development that led us down that yellow brick road could only have taken place with language. Schizophrenics have returned to that primitive brain structure and are probably incapable of this level of true happiness and often suffer from lack of joy (anhedonia). In fact, as I have asserted, when you talk to a schizophrenic, you are talking to someone with a prehistoric brain structure. If an anthropologist ever said to himself, "I wish I could talk to a prehistoric person," they need only talk to a schizophrenic . . . with one difference. A schizophrenic has learned to use language to express themselves (and nonetheless reveal the fascinating content of their mind and their manner of thought). Mr. Neanderthal or Denisovan could not do that. This is a valuable talent.

And what did humans do with this new linguistic property? You'll remember the consequences of contemplative thought. Remember those acronyms BDNF and VEGF? Those brain chemicals are mind fertilizers that arise in response to using one's brain as in thinking. They cause the brain connectors, the dendrites, to expand and multiply connections within the brain. So the more you think the better off you are, and they are euphoriants! One of the primary mechanisms of the new wonder-med ketamine is to stimulate BDNF and VEGF. It has an immediate, antidepressant effect. It is now approved by the FDA in a nasal spray yet as of this writing it is not widely available. The point being that thinking, contemplation, breeds happiness through internal brain mechanisms. College campuses are brimming with eager, joyous, hardworking students, all using their brains. It was not until contemplation became an option, which was not until language, that this starlet of euphoria was available to humankind.

What we also did with our spiffy new brains was vanquish our predators. We no longer have to carry around spears and clubs to protect ourselves, which were, to say the least, not always very efficacious. While there remains hunger in the world, in most modern countries there is a basic expectation of food availability which for the first time minimizes the necessity of expending energy to gain energy. This relieves a massive daily burden on us humans. Mr. Caveperson woke up hungry, and having no 7-Eleven to visit, he went out and scouted around for any kind of nourishment he could muster while at the same time exposing himself to predatory danger. Want a house? Go find a suitable cave. Want clothing? Go skin an animal and fashion that hide into a coat. Want shoes? You get the idea.

Can happiness really arise in those situations of daily need? No. A certain degree of leisure time is a part of the basic happiness quotient. To be under the constant pummel of mortal attack, wondering where your next meal might come from, how to clothe and feed yourself and your family, how to stay warm, and so forth, inhibits the calm repose of joy. There's no time or room for happiness, which arises out of a basic modicum of leisure. One could assert that free time is a modern invention also.

Now that we've covered thinking as a necessary precursor to happiness and free time with its freedom from predation, what else could bolster our assertion that happiness is a modern invention? Well, sex. Cavepersons had sex, of course, otherwise, end of humanity. So what I'm referring to is the difference between what one might call breeding and recreational sex. Without a modicum of free time, the latter would not exist. Mr. Neanderthal didn't have the luxury of lounging around having orgies. We do if we so choose to partake. So it was not until the language module delivered to us contemplation, which delivered to us problem solving, which delivered to us predatory relief, which delivered to us free time that we learned to exploit sex as a pastime. One might argue that this is a distortion of evolution's path, with evolution considering procreation the ultimate fulfillment of the genetic dictum, which we are all alive for. We exist to forward our genetic code into the future. Once that is done, evolution wants nothing to do with us. If you get a disease later in life when you are past reproductive age, evolution is wildly indifferent. Evolution couldn't care less if you have Parkinson's or senility; your job is done. Death, being part of this teeming tsunami of protoplasm, is a natural step in the evolutionary process. Once evolution is done with you, *sayonara*, you are free to *amscray*. Every judgment evolution makes about you, with its fickle thumbs up or down, the Siskel and Ebert of life forces called natural selection or later sexual selection, has to do with your survival and then ultimate reproductive fitness. Once we have fulfilled our evolutionary mission, we are disposable.

But beyond the issue of leisure time, there is the issue of sexuality itself. Sexuality, Freud theorized, has various components leading to mature configuration. Those components include oral, anal, phallic, and genital phases. The primordial caveperson would have gone through these stages rapidly, probably much more rapidly than do modern-day humans. We take a good fifteen years or more to reach mature sexual development. This drawn-out metamorphosis is a potential breeding ground so to speak for various sexual deviations or alternatives that may linger. Freud viewed fetishes as an expression of premature sexual longings that also may have entropic value. So this is another sense in which we expand the influence of sexuality in our lives and work them through or alternatively get stuck on one plateau or another. The point being made is that another element of happiness in regard to sexuality is the expanded time it takes to work through these theoretical sexual stages as we evolve into our mature sexual selves—another luxury our Neanderthal couldn't afford. One could describe this as an element of happiness only available to the modern human. Evolution trumps entropy with mature sexuality, but entropy intrudes on evolution with fetishistic, premature sexuality.

(Is orgasm not a state of entropic reward, the level of internal energy ultimately reaching a valley?)

But leaving sex behind, there is the issue of civilization. As mankind learned to neutralize predators, the advantages of communal living became obvious to us and to our ever observing Madam Evolution. As civilizations coalesced, the necessity for the golden rule was made bluntly obvious. To reap the survival benefits of societies we had to learn to coexist, and this coexistence depended on many disparate things. Impulse control became a highly valued commodity, and we learned to get along in a way that was dependent on an internal adjustment of values and complexity. It took a more highly complex brain, a result of the language module, to be able to navigate life in society and for us to don an effortless social strategy with which to confront the world. Society demands compromise, good-natured flexibility, and dopamine suppression. The Freudian schema assumes several hurdles that humans need to confront, like the Oedipus complex. It became unacceptable to have sex with one's parents or siblings, and this pseudo-evolutionary impeccability makes it clear that incestuous procreation can result in offspring that are potentially defective, an outcome that would be evolutionarily demurred. The fitness impairment and lower fecundity ratio (rate of reproduction) would rapidly cause extinction of such children. So in a sense, evolution found a way to preempt a doomed race by this Freudian imposition. (Evolution could not in and of itself defeat the behavior of incest, for which there is no gene. Humans acted, so to speak, as evolution's surrogate.) Patricide and fratricide also became taboos that man had to grudgingly internalize. These prohibitions were adopted with the ongoing demand of civilized living, and the ability to gate out unwanted impulses, in some cases, Freud said, led to neuroses.

One might conclude then that major changes requiring mental flexibility result in a number of dispossessed. It was language with its complexification of the brain that ultimately left 1 percent of the population, the schizophrenics, behind, and it was the transition to structured civilizations with its demand for internal prohibitions and metapsychology à la Freud that created neurosis. One could define neurosis as a byproduct of a nongenetic evolutionary mandate enacted by humans in the light of civilization's stentorian requirements. Freud uncovered this disease and analyzed the hell out of it. He suggested, as he stared out over the archeological artifacts on his desk, that a burgeoning of forbidden impulses could result in compromise symptom gratifications. These gratifications required a regression to childlike logic, the same limping logic we see in our primitive organization. In summary, neurosis would be the inflection of civilization's demand for internal regulation and ultimately repression of taboos, using childlike logic as a means of disguised gratification. Schizophrenia is the result of language's profound impact on the Homo sapiens brain, leaving some to regress to the previous six-million-year-old clapboard organization. Both illnesses then represent a return of the primitive and a victory for entropy, our fox-like, instant denigrator. Whenever a major brain structural change is demanded by evolution, some will be dispossessed. What will the next brain contortion be? And who will be left behind?

This also led to one other Freudian concept, that civilized acre of ground we call the *superego*. This little module rewards us with positive feelings if we do something altruistic. When we hold the door for the old lady behind us, we get a good vibe. The ability to go to work and interact on a social, nonviolent level rewards us with a sense of positivity and demands the kind of internal complexification only available to postmodern, languageified sapiens. This type of happiness is not available to children and schizophrenics and was not available to primitive people who had not achieved an adult level of brain organization. Socialization is a learned trait that school children hone for years but it coalesces around the cognitive maturation that is its prerequisite. The very recent installation of the language module made this possible and with it a facet of happiness previously out of reach to our grumbly, primitive, Neanderthal forebears.

For now I believe I've proven my point that happiness is a modern invention. Let's recap. Prehistoric humans were too busy scraping through a momentary survival to achieve happiness. The daily claw to the top did not allow it as leisure time, a product of civilization, is a modern achievement. The joy of children is a superficial happiness as they lack the adult brain structure to supply the depth of arduous glee adults possess.

Depression and suicide (or the loss of modern happiness) are modern inventions displaying a return to the primitive dread module with a state of hibernation, suicide being entropy's ultimate gratification, the return to inorganic matter. The ability to think that beneficently gifts our brains with chemical fertilizers is a necessary component of happiness that was unavailable to prehistoric humans lacking the language module. The superego is a facet of modern life perfunctory in children and possibly schizophrenics although they are very gentle people. This requisite cognitive complexity has been rescinded by evolution and the thief entropy. Free time led to recreational sex, a facet of happiness unavailable to Mr. Neanderthal.

We need to understand the purpose of organic matter in general. Such an oddball substance missing in the vast echoing majority of the universe must have a specific function or it would not exist at all. Our friend, inorganic matter, to which we all must eventually succumb, obeys the laws of natural physics. It doesn't challenge the old dictums and is duly rewarded with eternity's nod. That boulder in your backyard will potentially be there a billion years from now. But organicity defies laws of nature such as gravity (think trees that leverage zillions of pounds of potential energy against entropy by hoisting weighty masses of wood into the air, just as we do with skyscrapers) and of course entropy, organicity with its intense complexity and energetic endowments and as realized in the human brain, this being the most anti-entropic substance in all the universe. Organicity's purpose then is just that—defiance of the laws of physics—and the price organic matter pays for this defiance is mortality. We occupy a sizzlingly short lifespan on earth. Entropy and gravity both work on us constantly to end our defiant existence, and organic matter's quest to defy those laws of nature facilitates our return into inorganicity's eternity. Thus the purpose of inorganic matter is, for some reason (and only in habitats that are exquisitely hospitable to it) to defy some of nature's own laws. And we only succeed in this quest for minuscule expanses of time's vessel, which is why evolution is so keen on having us reproduce for otherwise we could vanish in an instant and need total re-creation.

I'll stop here, confident I've proven my point to the satisfaction of one and all. But wait, what can we learn from this that might help us prevent suicides? With the invention of happiness came the possibility of losing it. That event leads some to suicide, the most entropy-gratifying event of all and a slight to evolution. The answer is, however, not to prevent the happiness that leaves them wanting when the arbitrary resurgence of a primitive dread or hibernation state coupled with dopamine de-suppression calls them backward in time. Let's look at it for what it is. Fortunately, I can say that with persistence over 90 percent can now be helped with the array of treatment options available and some tenacity. Depression robs its victims of the latter trait. We must try to instill it in them.

Chapter 9

Alzheimer's and Parkinson's, Stuttering and Tourette's Syndrome, OCD, the New York Marathon, and Some Other Theories

Alzheimer's disease is being investigated with much vigor as our population ages. It differs from the usual accrual of senility that dogs old age. The brain deteriorates on a grand scale with profound memory difficulties and loss of function generally resulting in an inexorable, downhill march to death. We now have some medications bragging apparent usefulness in mollifying the decline, but they don't dramatically improve outcomes. What can our theories say about this disorder if anything?

Alzheimer's disease is a state of brain deterioration with terms like "neurofibrillatory tangles" and "amyloid plaque accumulation." Acetylcholine, the other neurotransmitter involved in motor control and probably thinking as well, goes into decline leaving a broad boulevard for dopamine to travel. As the brain deteriorates, this rumbling surge represents a coup by the primitive organization. This late-life storm cuts off stimulation of the higher brain centers, promoting even more deterioration, now with a nagging psychosis. A one-two punch, whatever process is damaging the senile brain is then augmented by the primitive organization's pouncing interference, with Alzheimer's patients often manifesting psychotic symptoms.

Parkinson's disease is by all definitions a dopamine disease. All attempts to improve its symptoms involve promotion of dopamine. It is, in a sense, the polar opposite of schizophrenia, occurring in late life, not early, and via a dopaminergic deficit, not excess. This late-life illness has some of the motor symptoms we see with antipsychotic side effects. This makes sense as the antipsychotics are blocking the hungry dopamine receptors. Could it be that the dopaminergic suppression of maturity is excessive, resulting in dopaminergic cell death? An intense clamp on dopamine might put the cells in a state of anticipation, which leads to their ultimate demise in later life. This situation may contribute to the emergence of Parkinson's disease symptoms accompanied by a dementing process akin to Alzheimer's. Dopamine-promoting treatments, including levo-carbi dopa (meaning dopamine), are an attempt to correct the deficit. Yet both Alzheimer's and Parkinson's display psychosis at times in fairly high frequency, and for these psychoses we use antipsychotics. (There is a black box warning not to use antipsychotics in the elderly demented patient as it may increase the risk of cardiac and vascular events slightly. However, having no other options, we often prescribe these medications in situations of dementia with psychosis. The risk-benefit analysis seems to justify this.)

My view of schizophrenia as a return of a primitive organization that is hyper-dopaminergic also implies that the process of adult cognitive development includes the suppression of dopamine. This suppression, as I pointed out, may be the defining factor that mankind has achieved in its advance from primitive to adult thinking. In fact, dopamine suppression may ultimately turn out to be the greatest achievement of mankind, originally birthed in the motor tract, the nigrostriatal. Once we began to utter words, this talent moved more centrally into the mesocortical and mesolimbic dopamine tracts, linking the motor function of speech with ideas. In order for us to emerge into a more adult cognitivity, this was necessary.

The ego, our friend Captain Kirk, helped us in an advance that was unrivaled in the protoplasmic world and evolution's giddy orchestration. Who could have predicted this turn of events? The use of language, simple symbol formation and all its accoutrements, skyrocketed our brains into reality testing. We became contemplating demigods, using this talent to ameliorate many of the world's problems (and sometimes create others), transforming our milieu largely in the direction of safety. Yet we increased the total excitation of our brains in a bold exchange of entropy for freedom from our dreaded predators. Agriculture freed us from the daily struggle for food. Yet we should keep in mind as well that it created caste systems and changed our diet to the point of protein deficiencies in some areas. A meat-based diet was necessary for humans to expand their brains, which rely on a food intake rich in fats. This could only happen once humans walked on two legs so they could hunt game (Lee and Yoon 2018).

What we are detecting is a seismic evolutionary failure in the arena of dopamine as a whole. If the lead neurotransmitter for millions of years in humans was dopamine, evolution's alteration required a readjustment of our relationship to this chemical behemoth to institute the mature cognitive structure typical of adult mankind as dopamine is held in check. Is this an integral part of the thinking process? What we are learning to do during our formative years is to reign in dopamine like a frisky calf, all of which unravels when schizophrenics break down, and to a lesser degree in the other mental illness diagnoses, and some may be especially prone to this undoing. In fact what might be inherited is a global potential for dopamine de-suppression increasing one's risk for any mental illness at all. This dopamine suppression may also be the difference between Freud's primary and secondary process thinking, the latter calling upon dopamine control, which gives us participation in our own thought processes. And the suppression of dopamine has the benefit of a gating function, blocking unwanted stimuli from the mind space. Thus the biochemical equivalent of Freud's concept of repression is likely dopamine suppression. Primary process thinking is akin to pre-verbal thinking but with words. The mind goes in its own, untamed direction without dopamine being controlled. We see in uni- and bipolar patients a return to the autopilot contemplation of primary process when they also dip into the raging dopaminergic stream, foregoing secondary process thinking.

It's easy, when contemplating something, to drift off into a daydream. Concentration difficulties are a primary symptom of both depression and mania. When one is dopamining, one has difficulty concentrating and is calling upon a chosen symptom of mental illness. One can learn not to do this as well simply by recognizing one's usual pattern of doing so. Whatever thinking our primitive ancestors did without language, it was on autopilot and lacked ego influence or reality testing. Yet the pre-verbal brain helped our ancestors survive the challenges that mother nature served up to them with the help of dopamine. Perceptual interpretation, memory, and associations all combined to promote survival. Evolution, learning from the mistake of dinosaur size, chose carefully. Cunning was its new mantra, and for that, a bigger and more differentiated brain was given the thumbs up. Evolution went from body size to brain function and won!

Tourette's syndrome is an interesting disorder. Sufferers may have involuntary motor tics ranging from mild to both severe and obvious. It is treated, if treatment is necessary, with antipsychotic medication. It is therefore a dopaminergic illness possibly localized in the nigrostriatal dopamine tract. It may go away toward the end of adolescence. There is often an element or hint of conscious control over the tics. Tourette's syndrome is related to OCD and has some obvious comparative similarities. Stuttering follows the same pattern, responding to dopamine blockers and may dissipate with age as can Tourette's syndrome.

If one conceptualizes Tourette's as a balance between dopamine and acetylcholine, clearly the dopaminergic blanket is too thick, forcing acetylcholine to muscle its way through in a tic. Then according to my hypothesis, dopamine is suppressed as we mature. The neurotransmitters are now more balanced, and acetylcholine has a thinner blanket to push through, no longer needing to pummel its way forward like a prizefighter. Acetylcholine represents the voluntary side of movements, but without the blanket of dopamine those movements would be uncontrolled. We see this in Parkinson's disease, a deficit of dopamine, and with heavy antipsychotic use in the form of what we call *extra-pyramidal side effects*. The latter are discomforting symptoms of motor restlessness, acute dystonias involving intense muscle spasms like back spasms or eye rolling or bicep contractions, all of which contribute to the refusal of patients to take their medication. The dopaminergic blanket is now too thin, allowing acetylcholine to burst through almost unchecked. Now we have to rebalance the neurotransmitters with medications that suppress acetylcholine.

There is a need for balance, and eventually dopamine recedes under the suppressive influence in early adulthood, sometimes resolving some of these problems like tics and stuttering. But these motor symptoms involve the nigrostriatal tract, and there are other tracts that dopamine suppression or lack thereof can wreak havoc with. ADD and restless leg syndrome may involve a lack of dopamine with an excess of acetylcholine, and we add dopaminergic enhancers to treat them. This restores the dopamine-acetylcholine balance in the other direction, relieving the symptoms.

So the muscular submergence of dopamine that got its start in the motor area of the brain gradually leached over to the mental arena. How did this happen? Well, one of the primary muscular assignments once we learned language was speech, involving dozens of muscles in the mouth, lips, tongue, and throat. These muscles, demanding as much suppression of dopamine as any other, also involved the conceptual system as the ego fumbled through the cluttered preconscious to decorate concepts with words. Speech then combined the motor task of oral enunciation with concepts. Thinking then borrowed the trait of dopamine suppression with its gating function and ego enhancement from the muscular system in speech.

We are now in a position to understand symptoms like delusions completely for the first time. Freud theorized the psychodynamic factors behind the choice of delusion. However, he did not explain the purpose of the delusion itself. The delusion, like all symptoms, can be seen as a means of supporting the de-suppression of dopamine. It is not the unleashing of dopamine that creates the symptom. Dopamine seeks a path to unsuppression and then drags in with it whatever symptom is viable. In one person it may be anxiety attacks, in another it may be obsessions. Delusions have nothing to do with their content except as they reflect the psychodynamics of their believer. Schreber's belief that his doctor was trying to kill him may very well have been due to a father complex as Freud suggested. It did allow Schreber, within his state of cognitive regression, to gratify his wish to occupy the body of a woman. But the purpose of a delusion rests not so much in its content as in its goal: to support the de-suppression of dopamine at the highest level possible. Why? Because in its flight from dopaminergic suppression, the mind seeks to restore entropy. The mind seeks a lower state of energy by giving up on its hold on dopamine. Whether it be anxiety, depression, poor concentration, Tourette's, schizophrenia (the ultimate dopaminergic unleashing), sleep, psychedelic use, or any other symptom, depending on which dopaminergic tract is involved, this rescues and promotes the de-suppression of dopamine. The process pulls the delusional concept in with it, which reinforces and champions it and gets the individual in touch with the primitive thinking module 1.0. This is why delusions are impervious to reality. They serve a biochemical purpose. The undoing of the repressive burden, whether by psychoanalysis or other means, frees dopamine for ultimate suppression, leading to greater mental stability and a higher resting state of mental energy.

But to complicate an already scrambled issue, there's another aspect to this complex symphony. At least for the nonschizophrenic, the symptoms fashion another entrenched and desperate ideal and that is to forestall a coup by the primitive organization. Given that there is constant pressure on the part of entropy to back flow to a less-organized state of mind, the primitive organization is able to beef up a threatening posture in certain people who resonate with its intentions. And in dueling with this entropic absolution, just as Schreber did when he accepted having been transformed into femininity, they strike a bargain with the process and offer up a sacrificial lamb in the form of a lesser homage to entropy. That homage, treacherous as it may be, engages one of the less dramatically entropic diagnoses than schizophrenia. They may develop the trappings of major depression, the repetitive corruption of OCD, or any number of symptomatic ills that avoid a complete takeover by the primitive organization. For example, symptoms like self-cutting reflect a means of avoiding the fury of schizophrenic wrath. Often we see schizophrenics given diagnoses other than that before their ultimate break down. These on-the-way-to-psychosis ills reflect an attempt to forestall the primitive organization's coup, which ultimately fails in their case but succeeds in the case of depressives or manics. This bargain with entropy mitigates at least a regression to its ultimate trapping, schizophrenia, with suicide being the only more radical expression of entropy's desire.

* * *

You're in a crowd of super-pumped, icy thin, world-class runners. It's the New York Marathon, and you're psyched, having trained for this for what seems like millions of years, and you know you're ready. At the very least you're going to make it to the finish line. After endless, breathless practicing in parks and streets and playgrounds, you've got what it takes to endure. You're genetically fit and phenotypically on the mark. You've got all the accoutrements, the Nike shoes, the Adidas hat, the sweats, the socks, and the shirt; you're ready to rumble. Standing there among thousands of people you feel small, facing the starting banner, hopping up and down, stretching your calves. Bang! The starter pistol goes off and you're on your way in a teeming sea of humanity surging around you.

The temperature is right but it's a little cloudy. Your body is moving, breathing comfortably, with no cramps or nagging, treacherous aches and pains. You're in the groove and make the first turn. It's all good, and you're running like a well-oiled machine. Staff on the sidelines offer you orange slices and bottled water, but you don't grab that, not yet. You're sweating but not so badly that you need replacement. Feels good, the endogenous opioids coursing through your veins like an addict's prayer.

Then suddenly there's a downpour, the sky punishing everyone with a drenching rain. Drops are bouncing off your cap like ping-pong balls, your sweat's mixed with water. It's a little chilly now, and you're thinking, "Let it stop; we don't need this added woe." But it doesn't. Mother Nature has decided to rain on your parade big time. People will get pneumonia, and they'll be carrying extra water weight, but that's the dicey marathon. You take what you get. Then an announcement comes over the loudspeakers: "Everyone must take the subway to the finish line." What? Yes, you heard it right, the subway. People keep running in disbelief, not knowing what to do. But the announcement keeps blaring: "All participants must take the subway to the finish line. No one who gets there on foot will be acknowledged."

Finally everyone starts searching for the nearest subway entrance. Scratching their head in disbelief, they decide that they'd better follow the rules no matter how cranky if they want to finish although this wasn't what they trained for. You do the same, following runners who are dribbling into subway entranceways, scrambling in obedience. They hurry down the stairs and line up at the ticket booth. It's drenching rain and the line goes all the way up the stairs and out onto the sidewalk. People are grousing and demanding tickets. You wait your turn patiently while the ticket machines are being stuffed with money. Others brought no cash and have no clue how to proceed. Even with tickets some people can't seem to figure out the turnstiles' mysteries. People are crowding the platform, waiting for a train to arrive. The platform's so crowded that runners are in danger of being pushed onto the tracks. The line is still up the stairs and onto the sidewalk. Finally a train comes but it's fairly stuffed and people have to be ushered off like royalty. Then there's a rush to squeeze into cars that are already half full. Race participants with their numbered vests and squishy Nikes are running up and down the platform trying to find an open car. The doors clamp shut and the train leaves, abandoning the desperate like you.

Finally you get to the ticket man who grumpily sells you a token. You walk up to the turnstile and clunk it in. Squeezing through, you see the crowd is bristling intensely on the platform. Everyone is waiting for the next train to arrive. Walking down a bit, you stand there hoping for salvation. Several others weren't lucky enough to have money to buy a token at all. When the train arrives, the mob is poised to push their way in. As soon as the train stops the doors creak open and a large blob of people are let out by the crowd. Then as soon as the departures are gone, runners burst in frantically. You push your way forward, moving as agilely as possible toward the door. You try not to let others cut in front of you but it's impossible with people jostling and elbowing in wired exaltation. Finally you squeeze ahead enough to pass through the doors. You're in! You hold on to a metal pole for stability. The doors close. Everyone breathes a sigh of relief, and you absentmindedly start staring at the subway map up on the wall. The stops are listed in order and you look for the one you're supposed to get to. Reading each stop, it's hard to find the golden rainbow you're looking for. The different lines branch out in all directions, one to Queens, one to Brooklyn. The train seems to be going fast and not stopping anywhere. It must be an express. You are lucky to have gotten on at all.

This is the situation schizophrenics find themselves in. When evolution made its herculean course correction, they were cash-poor for a token or got elbowed out and now, at perhaps age twenty through no fault of their own, a pre-subway state of mind has taken over and recaptured them, leaving them vacant at the turnstile. The marathon, representing evolution, made an egregious hairpin turn, and some were inevitably left behind with a departure of that magnitude. This explanation for schizophrenia answers many of the basic questions we have about the disease. And we're all still caught up in this marathon rule change, which will be ongoing for thousands of years more after which we'll all be on the subway and mental illness will be a thing of our antique past. Far enough from the gluey primitive, entropy will lose its seductive lure.

A key question and the great paradox of schizophrenia is, Why is the disease not extinct? As we've discussed, it's a perfect candidate for extinction since natural selection's fickle finger of fate is no friend to an illness that reduces your ability to survive, compete, grasp complex concepts at an adult level of rules, and so forth. Similarly, sexual selection gives the thumbs down to infirmities that arrive just as the sufferer is approaching reproductive age, which reduces his or her desirability as a mate. The schizophrenia fecundity ratio, the rate of offspring production of schizophrenics compared to the normal person, is far less than average. Such an illness would, according to Darwinian principles—sexual selection or natural selection—vanish in the breeze. Many illnesses that we've never heard of no doubt became extinct when they did not meet Momma Evolution's survival criteria. Conditions that are clear Darwinian losers must have some explanation if they manage to survive. For example, Down syndrome is a spontaneous mutation in early embryonic development, trisomy 21. It happens by chance and is created de novo each time, not passed down from one generation to another. Diseases that arise spontaneously as such can survive ad infinitum since the cause is a genetic glitch that does not require ongoing passage from one generation to the next and thus is not governed by evolution. An illness like gout, which is sometimes caused by excessive drinking of alcohol (although it has other etiologies) can continue as long as the poison that creates it seduces its tipsy victims. Illnesses that result from infection may occur at any time if an organism is around to infect. For example, appendicitis can occur in anyone with an appendix (a useless holdover organ from our past) if it gets infected by a bacteria. Only mutational illnesses, those caused by gene errors that trickle from generation to generation, are ruled on by Darwinian principles that govern any mutational experiment. The intense search for a schizophrenia gene seems unlikely to succeed, and in fact has born no fruit. There are single nucleotide polymorphisms (gene abnormalities) that are more related to schizophrenics than nonschizophrenics. But polygenic risk scores, derived from such groupings, are not specific to schizophrenia and have little predictive value regarding who will or won't ultimately become schizophrenic. In fact, if anything they point to a universal causality such as the tendency to de-suppress dopamine as being genetically endowed. This global vulnerability would explain the multiplicity of diagnoses often seen in families of schizophrenics. Years of intensive search have not uncovered one gene responsible. But a theory that states that schizophrenia is an evolutionary event, not a genetic one, is perfectly consistent with the facts. Genetics is a highly complex, emotional issue. I am no geneticist. But there comes a time when we should concede that we are not going to find the genetic basis of schizophrenia. An evolutionary event also accounts for the anthropo-parity principle, that is, the steady, uniform rate of schizophrenia worldwide. One percent will not get on the subway until the course correction is done, the rule change is complete, entropy loses its grip, and primitivity holds no sway.

Examine any category of person or phenotype—for example, people who can bowl 300. Those people are going to have certain genetic differences, genotypes, and therefore traits or phenotypes associated with them. We might find that their thumbs are shaped slightly differently. Perhaps they have excellent hand-eye coordination. The movement of a large ball down an alley may hold particular interest for them. But environmental factors also enter into it. Their fathers always looked up to good bowlers. Their grandfather was a world-class bocci ball player. Their spouse or partner is very impressed with bowlers. But those variations don't cause 300 bowling. There is no perfect-bowling-score gene. They contribute to it, or in the case of an illness, shift the odds. But in the case of schizophrenia, it is not those characteristic vulnerabilities that are the disease itself. A brand new evolutionary process that is still enriching wrought largely by language has created this gnarly situation. The good news is that some time hence, oh perhaps ten thousand years, the situation will go away as this change in paradigm reaches completion.

Evolution is a vast, teaming swarm of protoplasmic change. Organic matter is spewed forth from the inorganic to participate in life and reproduction then swiftly return to its inorganic roots. Evolution wants nothing more than to facilitate organic matter through time, and that facilitation is governed by Darwin's architectural rules—natural selection initially. Living matter's metamorphosis was totally governed by its ability to survive. Mother Nature created the punchy habitat in which living beings existed and the experiments that moved them forward called *mutations*. She or something created DNA and deserves many kudos for doing so. (DNA's most salient characteristic may be that it makes use of entropy's talent for disorganization in the form of mutations, thus trumping entropy's desires.) DNA has the capacity to mutate, screwups happen, and these mistakes were experiments in survival. Natural selection ruled up or down on these modular DNA creations, and the mutational changes in the face of adversity either helped the being survive or dug potholes in its path. Protoplasm advanced by these rules, and the brain of man enlarged simply because its increased size increased survival. Once language appeared, all Hell broke loose, intensifying our complexity and energy, kicking entropy in the derrière, the mind being the most anti-entropic substance in all the universe. Whether fast or slow, the changes wrought by language on the Homo sapiens' brain were startling. The brain lateralized to accommodate language, the left side swelling to house language's demands and the skull enlarging to hold it. Humans moved from experiential-physical beings to conceptual as our brains began to solve the most urgent problems of survival. The benefits of civilization, education, religion, and morality all accrued in conjunction with the neuronal razzle-dazzle. We turned our back on grumpy Mother Nature and paved the earth, developing an expectation of survival. Neurotransmitter changes in the brain with principally dopaminergic suppression were required of this new, improved, human-mind experiment.

And as you can see we are still in the very inception of all this change, unable to notice it as it creaks by over generations. Perhaps its most salient characteristic is a move away from primitivity and entropy although we're still close enough to its gnarly influence for it to recapture 1 percent of the population worldwide. The brain ballooned on one side and granted us the passion of adult thinking to navigate a complex symphony of laws and human interactions. All of this accrued toward safety and freedom from the grotesque Addams Family of predators that the Neanderthal was subject to as we no longer expected earth to reclaim us in our twenties and thirties but to scrape by now well into our eighties or beyond. We neutralized through contemplation, with weapon management, cooperation, and ingenuity, most of our scariest adversaries except man himself. We are still in the midst of neutralizing predatory illnesses, predatory nations, predatory deficiencies of food and water, and predatory weather, and standing within this dinosaur footprint, not everyone is comfortable. The primitive, seeking to reassert itself, does so by de-suppression of dopamine in 1 percent of the population as regards schizophrenia, perhaps the most egregious example of return to our entropic past, but other mental illnesses in far greater numbers also bear the watermark of primitivity. The more our brains expand on the left side, the farther we get from the primitive. Yet we are not so divorced from that recent change that it does not rise up in some of us and exact its toll aided by entropy. Primitivity lowers brain excitation, and organization, and therein lies its inchoate draw. Entropy is also a seducing factor in multiple attractions like drug use, TV watching, art, sexuality, and other transformative distractions that lower overall excitation of the highly complex wiring that is our brain, the universe's most antropic excrescence.

The age of onset of schizophrenia is most likely a function of the primitive organization's need to reach a basic neuronal maturity as we grow up just as the higher, more recent centers of the brain need to do so. Once that is symphonically blessed, the primitive organization is ready to make its corrupt invasion. But also, as the balloon of repressed material swells it requires greater and greater effort, possibly reaching a breaking point as maturation proceeds. We may reach a point in our brain development beyond which this organization is effectively locked out through dopaminergic suppression or some other means. It chooses a point just before full brain maturation to strike, dragging its victims backward to prehistoric mentality. This point of no return happens around age twenty-five, after which schizophrenia is extremely unlikely with an incidence rate of close to zero.

Once the primitive organization sets in, the higher structures of the brain that had been gearing up are short-circuited, sparks flying like spliced battery cable. They go into a state of sleep-like sensory deprivation and are not being used in the manner they've become accustomed to, are no longer receiving the voltage expected. Just like any muscle, brain tissue needs input or it will wither and perish. We make sure that paralyzed muscles are exercised by others to remain active. As in sleep when the brain is sensorily deprived creating dreams, it sends out self-stimulatory impulses in the form of hallucinations, auditory, visual, or tactile, and since there is abundant visual input, most of the hallucinations are auditory as opposed to the repose of sleep when our eyes are closed rendering dreams largely visual. Unfortunately these hallucinatory impulses are taken over by the new world order, the primitive organization, and used against the individual and especially their ego in gruff demands of self-destruction or pejorative rejoinders, giving voice to entropy's wish to have us return to the inorganic. The ego stands like a lonely clock tower against the primitive organization and is in fact anti-primitive, having delivered us from the gnarly grasp of infantile incantation in the first place. We can categorically sum up mental illness as a return of the primitive, a wallowing in entropy and backslide of dopamine suppression. Entropy is on the primitive's side, and loss of dopamine's suppressive gifts carries with it a loss of gating function that blocks out unwanted stimuli from our blessed theater of contemplation. So when schizophrenics collapse in a tsunami of dopaminergic backflow, gale-force intrusions of input saturate their brain space in the form of whispering incantations, howling critical voices, mumblings, and blaring vocals.

All of the above leads us to the question Could there be one common cause for all mental illnesses, one radical vulnerability in all its tentacular variation? This question has plagued theorists for centuries. In the *Diagnostic and Statistical Manual of Mental Disorders* (DSM-V), the American Psychiatric Association's handbook of diagnoses, there are dozens of precisely enumerated diagnostic categories. Each one is defined with a number and a list of symptoms. These symptoms or criteria must be present for a diagnosis to be made. While being a worthy attempt at accuracy in diagnosis, it also has little elaboration of anything like causality, nuance, or reverberation. And if a symptom or criteria's not cited, an insurance company may decline to reimburse for treatment of something that the APA does not acknowledge. The fact remains that many recipes with their diagnostic ingredients are listed as illnesses of the mind. Could there be a common etiology, a universal core for these problems? In ages past, hysteria was blamed on a wandering uterus and mental illnesses were ascribed to various neurohumors or substances that fogged and distorted the mind. We still look to infectious agents, concussions, and other brain insults as the etiology or cause for atypical thinking and behavior. If one etiology exists, it is the de-suppression of dopamine and the individual's vulnerability to this that equals the magic artery of explanation, the bankroll of likely possibility. After all, we see families with a hodgepodge of diagnoses lined up without any focus as if the tendency to mental infirmity as a whole blossoms indiscriminately. Polygenic risk scores not only predict one diagnosis like for example schizophrenia but the whole list of major mental disabilities. Each one of us inherits a score, a tendency-to-dopamine-de-suppression number, that predicts our likelihood of blurring into mental demise. This would explain what we see statistically: an increase in vulnerability to Dad's addictions or Mom's panic but with our own variation on the theme and of course with the addition of a nongenetic contributor, evolution's recent shapeshifting.

Maturation is a like the Sisyphus myth. The dopaminergic boulder is rolled uphill into a highly placed suppressive crater, there to rest in lifetime repose far from entropy's demeanor. This well-placed boulder typically succeeds in posing no threat to our brain. The elucidated benefits of this resting state is the adult complexity of mind earned over the past fifty thousand years of speech, a bulwark of hefty ego function, and brother dopamine's gating out of intrusive distractions, an assignment that takes constant energy. The risk is a bitter dislodgment of place wherein our boulder careens down again, crashing into the lowest level of energy and complexity, entropy's home, Mother Earth's floor. Thus schizophrenics find their boulder barreling down that iconic hill in full disobedience of their wishes, back to the prehistoric position of our Neanderthal and Denisovan forebears, a tribute to dopamine's de-suppression.

But in the mood disorders, uni- and bipolar, we see a return to more primitive, primary process ruminating without the loss of adult cognitive rules. Major depression is a return to a primitive hibernation state without the wholesale collapse in logical processes that we see in schizophrenia. It shifts the usual thought pattern from secondary to primary process thinking, the embattled autopilot of the past six million years or so. If happiness is a modern invention, depressives return to the affective state of the hibernating cave dweller. Mania, on the other hand, is a desperate flight from dreaded depression and encapsulates the level of primitivity imposed by it. It's as if the boulder rolls only halfway down the hill, lodging on a balcony. Again, primitivity embraces entropy, a return to a higher level of disorganization and lower excitation than is found in the highly structured, razor sharp, hierarchically organized modern mind. Dopamine un-suppression is a key in this boulder-rolling analogy with entropy supplying the gravitational pull.

Freud's neurotic patients displayed primitive, metaphorical thinking in their symptoms as well. These symptoms were compromises between taboo impulses—impulses that became undesirable as civilizations accrued—and their gratifications. For example, if a woman entertained a hysterical pregnancy as a means of gratifying a wish to have sex with her father, she was certainly using the same sort of childlike thinking that primitives engage in when opening windows and doors in a mistaken attempt to assist childbirth. Elements of primitivity sneak into all mental illnesses one way or another. The question then becomes is this a result of evolution's course correction and the unique demands it made on our central nervous system?

My answer of course is yes. Without evolution's metamorphosis we would not be seeing either schizophrenia or any other mental abnormality.

This leads us to ask what is normalcy, another age-old conundrum. Each person's normalcy is different. But by the definition outlined above, normalcy would be the safest distance from primitivity, a higher level of mental organization and excitation 2.0 with smooth-working structures hierarchically organized and defying entropy's desires, and the most intense level of dopamine suppression, the boulder nestled soundly at the hilltop. Mental health necessitates successful suppression of dopamine, perhaps Homo sapiens' greatest accomplishment, and highly attained language skills with adequate brain lateralization orchestrated in a grand mental symphony. Freud expressed it much more simply by saying that if one can work, and love, and be reasonably happy in life, that is what one can expect. I would have to agree with that commonsense assessment. In order to understand who we are, we must take into account our lengthy past and the modifications wrought by the change in evolution's blustering trajectory largely due to language.

* * *

In the time I've been writing this book, many things have changed in my life. During all this it occurred to me that I have not given the reader a sense of what psychiatrists do day to day, a crucial diorama for people to glimpse including those who are thinking of seeing one or becoming one or who are just curious. Perhaps you have a relative who has seen a psychiatrist. Huge numbers of people do see them. What exactly does a psychiatrist do every day?

I wish I could tell you it goes like this: we open a patient's chart, review their symptoms, note the lab tests that prove or disprove exactly what their illness is, meet with the patient, prescribe the perfect medication for them and their problem, hand them a prescription, and then see them again a few weeks later happy and basically cured.

Um, nope.

In psychiatry we have always looked for a blood test that would reveal the diagnosis, or a CT scan or a . . . you name it. We do not have it as yet and remain in a fairly primitive state of knowledge despite all the fancy medications and diagnostic categories. What we have is clinical diagnoses based on symptoms, history, response to previous treatments, and so forth, and no lab test, X-ray, or genetic locus to implicate in the disease process. We would love to have a test that proves definitely that a given patient has a given disease. Instead we rely on what the patient tells us and what we know of their past from documents, recent and dated, and family reportage. Unfortunately, patients aren't always in a position to fill us in accurately on their symptoms and treatments. And even if they are, we must rely on our judgment and experience to match their haphazard narrative about themselves to the templates we've secreted in our mind's eye to arrive at a decision about what they have. Nor is it common for patients to have just one wistful, isolated diagnosis. More often they have a combination of issues and diagnoses that cross categories and make designation tougher. Even if we are convinced that a patient has, for example, bipolar disorder and PTSD, there may be a gamut of symptoms to address. Which is most important, their mania, their endless nightmares, their depressive suicidality? We attempt to treat it all as best as we can. Often there is some trial and error in finding a medication or more likely multiple medications that might make their life more bearable and stable. We would love to find just one medication to cover all the symptoms, yet polypharmacy, using more than one medication, is often urgently necessary. It's hard to stabilize a patient's mania if nightmares from PTSD won't allow them to sleep. So we try to treat a broad splatter of symptoms as best as we can. The truth is that a medication that works for one patient may not work well at all for another or may have intolerable side effects for a third. We rely on our experience to use medications we've had success with in the past, but groping trial and error is an integral part of treatment as is a grudging tenacity. Different psychiatrists will treat the same patient differently with the same goals, and all roads may lead to Rome. The goal is stability, safety, freedom from debilitating symptoms, and functional competence in a framework of fluctuating contentment. Medication is an art more than a science and we all struggle with this, resorting at times to nonmedication treatments like ECT, transcranial magnetic stimulation, and talk therapy, for example, all of which play a role in the total recovery of some patients. We are lucky to have multiple effective options.

* * *

I am not the first person to theorize about evolution and schizophrenia. Others have thought to do so and made some important points about it. You deserve to hear from some of them so I will try to present a straightforward, unbiased account of their ideas. The theories tend to fall into two broad categories: those that portray schizophrenia as a natural selection advantage, and those that portray it, as I have, as a disadvantage and find ways to explain its persistence. Balance theories attempt to balance the obvious drawbacks with the assumed benefits of schizophrenia evolutionarily. Byproduct theories assert that schizophrenia is a byproduct of something else but they tend to cling to genetics as the basic etiology. Most theorists agree that schizophrenics reproduce at a lower rate than the average individual, which is statistically undeniable. The anthropo-parity principle, that schizophrenia has a steady 1 percent incidence in all places, must also be explained by whatever theory one espouses.

Some theories attempt to rationalize the persistence of schizophrenia as a result of some cryptic reproductive advantage it might give the group of relatives around the schizophrenic person. This is not borne out by research. Those relatives may carry some vague schizophrenic genetic tendency without actually bearing the burden of the disease. Some studies have declared that relatives of schizophrenics have superior academic success. However other studies highlight lower IQ scores among relatives of schizophrenics. It seems unlikely that the relatives of schizophrenics will have a wide enough panoply of advantages (if any) to keep a steady 1 percent rate of the disease alive even if one believes in the genetic transmission of schizophrenia. Yet this is an interesting theory that attempts to hobble around the genetic problem.

Byproduct theories assert that schizophrenia is a secondary result of some other event. My theory would fall into this category. Crow (2006, 2011, 2000, 2012) suggests schizophrenia is a byproduct of evolution but supports a genetic origin, which according to Darwin is not logical. He makes note of a possible decrease in cerebral laterality in schizophrenics. Laterality, as we have discussed, is a harbinger of language acquisition.

Crow seems to be suggesting a failure of language-ness, which is not necessarily the case in schizophrenics. In fact they achieve adult communicative function before the onset of their illness. To use the Sisyphus analogy, they role the boulder up the maturational hill, but just before it is lodged safely in the crater of adult functioning, it rolls back down and nestles wanly at the entropic bottom. This may imply that the rolling up the hill confers some permanent change or fixity in the higher centers of the brain. The ego's conceptual input to the forebrain becomes its modus operandi, and once the boulder rolls down again, those brain loci no longer receive the stimulation they need and so begin to self-stimulate in a desperate drive to sidestep atrophy. Crow seems to believe that a "speciation event" that distinguished humans from apes was linked gene-wise to language as well as the sex chromosome. Crow also seems to believe in a continuum model of mental illness such that all psychiatric problems fall somewhere on a related trajectory. I tend to agree with this. Using the Sisyphus model again, uni- and bipolar illnesses may represent a boulder that rolls halfway down the hill accompanied by a de-suppression of dopamine. This happens despite the boulder finding a place on the summit. Still there are implications in this partial roll-back that could be viewed as a last-ditch effort to forestall the boulder's nonstop roll back to entropy's valley. In fact, all mental illnesses except schizophrenia might be a nail-scratching effort to re-channel the torrid stream of de-suppressive dopamine away from the ultimate schizophrenic cornerstone.

And if rolling that boulder up the hill does cause a permanent change in higher structural brain functioning, a modern sitcom of happiness may be a result of that change, making happiness truly a recent invention. In short, as humans advanced in cognitive ability, they laid a foundation for happiness coinciding with dopamine's punctate suppression, relational ability, civilization, laws, religion, and so forth. Sexuality also changes with modernity in that it is more refined, more orgiastically pleasure driven than just reproductive. (Evolution made a sneaky bargain with entropy in that orgasm is rewarded with a relaxed, deflated state of de-excitation. Fetishes, on the other hand, seek that entropic nirvana without achieving insemination, sexuality's highest evolutionary mantra.) Of course thinking itself could only have arisen after language when we realized that we could talk inwardly to ourselves. It promotes BDNF as does ketamine, strengthens the ego, and complexifies the mind in a way no prior activity could, and of course only for us lucky Homo sapiens. So true happiness involves cognitive maturity, dopamine suppression, leisure time, social adeptness, the ability to conceptualize, superego, sexual maturity, social strategizing, and freedom from adversity—all modern accoutrements gathered from the well of language.

Sleep, a useful comparison to schizophrenia in that it grooves a path back to the prehistory in us, engulfs the mind in night-time hallucinations called *dreams*. The mist of darkness and quiet swirls around bringing us back to the state of entropic occupation of our forebears. Indeed, sleep is an entropic dance, restorative in its anergic repose. This black state of deprivation seduces the forebrain into dreamy self-stimulation, protecting it from a total absence of input and offering a movie screen to replay the day's events in creative restitution. Here the primitive organization holds full sway as the rules of dream logic melt into childish regression complete with the backflow of dopamine we've come to anticipate, the same urgent backflow we find in an acrid cup of ayahuasca or the collapse of the mind into schizophrenia's tentacular crush. The opposite of sleep, thinking, shoves dopamine into its suppressive corner and reverses the regressive urge.

Polimeni et al. (2003) states, "Most evolutionary based hypotheses that are related to schizophrenia accept the assumptions of the schizophrenia paradox." This implies that schizophrenia should have been extinct years ago. The authors note that schizophrenia is considered a high-prevalence condition as 1 percent exceeds common mutation rates. This endorses the assumption that schizophrenia is not a genetic event but an evolutionary byproduct destined to unwind into the future. Genetics may follow such events in that a phenotype may be associated with several risk alleles: that is, genes that are less common in the normal population but have secondary or tertiary connections to an identified trait by, for example, increasing one's risk for mental illness as a whole—a sort of global risk score or dopamine de-suppression score.

Models that attempt to assert a natural selective advantage to schizophrenia have a difficult trail to blaze. For example, some suggest that the reduced fecundity in schizophrenia is balanced by some ill-defined higher resistance to certain infections or other deleterious incidents. This balance theory seems to be a stretch of credibility with little to no concrete evidence. Polimeni et al. (2003) proposes an advantage to schizophrenics in ancient societies that saw them as shamans or religious leaders because they could talk to spirits. Their psychosis would be viewed as a special gift used by the group for guidance. This theory seems somewhat dubious given the schizophrenic's difficulty adapting to social situations let alone a leadership role. Nor would it necessarily confer the likelihood of genetic survival. Even if one leader or shaman succeeds in procreating, the rest would not be so blessed, nor does this explain the ongoing rate of schizophrenia today when they do not play such a role in society. While examples of this type of individual might be Jim Jones or Charles Manson, those are luckily few in numbers. Others have suggested that schizophrenic qualities such as paranoia produce charismatic leaders such as Joan of Arc or Hitler but there is no evidence to substantiate either as a schizophrenic. Paranoia renders most schizophrenics much too wary of any group to hold a leadership role, more likely leading them to a hermit-like existence far from human interference.

Schizophrenics have been at times considered mad geniuses with implied selective advantages. This is a mistaken notion, with high-IQ schizophrenics being just as rare as high-IQ nonschizophrenics although there is undoubtedly a group of high-functioning schizophrenics. Also ascribed to schizophrenics occasionally is an increase in creativity. While this may be true of some, it is hard to translate it into a global reproductive advantage. Others have proposed an advantageous territorial instinct in schizophrenics presumably springing from their low tolerance for stimulation. But territorial instincts are innate in humans and lead to conflicts large and small. Others have postulated increased resistance to viral infections in schizophrenia with little if any corroborating evidence.

Australian Aboriginals became effectively isolated from the rest of humankind about sixty thousand years ago and have been shown to have schizophrenics in their population. Why would this be if schizophrenia evolved within the last fifty thousand years? The key issue is language, its cognitive enhancement in the society, and how the society has evolved in the recent past. This is the approximate time frame within which language surfaced in its most complex form. Aboriginals, primitive as they may be, are not living the lifestyle of Mr. Caveperson. They have language and an advanced societal structure that places them in a post-verbal environment. It is the shift from pre- to post-verbal that creates the evolutionary course correction that places 1 percent of the population at risk for schizophrenia.

The persistence of schizophrenia tells us that it is not a genetic event but rather an evolutionary one. Theories that attempt to rationalize it as a genetic disease are on shaky ground from day one. It is true that there is a slightly higher risk of schizophrenia in children conceived by older parents. (By the way it is also true for autism and bipolar disorder that paternal age increases risk.) This is taken as proof that some mutation must be involved since age does increase the rate of mutational events in sperm that succumb to entropy's influence. However there is no mutation that accounts for schizophrenia. There are clusters of genetic abnormalities that are more often seen in schizophrenics than others. Yet a reduction in reproductive fitness by as little as 0.003 percent has been shown to work toward the extinction of an illness (Uher 2009, 2014).

To present another example, if we take a group of people, let's say those hit by lightning during the course of their lifetime, one will find clusters of genetic abnormalities that are more likely in them. There may be genes that have a subtle influence, increasing their risk for lightning strikes. These genes may work to encourage their outdoor time, their tendency to stand in areas where lightning strikes are more common, their lack of fear of thunderstorms, and the like. These genetic differences may add up to increased risk yet that is not the cause of their injury. Lightning is. The genetic variants with phenotypic impact that are more likely in schizophrenia may have an indirect connection to its appearance but that does not make them the root cause of the illness. Rather it is a resurgence of the primitive organization, so recently overcome according to my theory, that existed millions of years ago and a failure to maintain the modern mental organization that exists today along with a de-suppression of dopamine and the indulgence of entropy that seduces us back to the prehistoric mental dance. This explanation is more plausible and comprehensive than any attempt to rationalize schizophrenia as a product of natural selective advantage.

While genes are considered by some to be the culprit in schizophrenia, environment is also considered by most to launch its unfolding. In fact some external insults will only have an effect if certain genes are present. There are polymorphisms that increase the risk of psychosis only if one uses marijuana for example or if one experienced cytomegalovirus in utero. These gene-environment interactions may play a significant role in the etiology of certain illnesses or a very insignificant one. They may also lead us to overestimate the concordance of schizophrenia in twin studies that fail to take into account the shared environment of the twins whether mono- or dizygotic. If nothing else, identical twins share a common experience and placenta in the womb. Insults in utero such as infections or trauma are felt by both twins, elevating their joint challenges even if they are adopted after birth or separated years later. A shared environment increases the risk of concordance without the influence of genetics. If the twins both have a global risk allele, for example, raising the odds of dopamine de-suppression as a whole, then an environmental insult will effect both. In fact, polygenic risk scores suggest an increased risk of mental illness as a whole lacking specificity for schizophrenia. Fraternal twins may differ as widely as any other siblings in respect to their risk alleles. Some authors have argued that twin studies have vastly overestimated the concordance rate of schizophrenia. Factors such as an urban upbringing, minority status, parental abuse, head injury, being born in the winter, and drug exposure or use may increase your risk of schizophrenia. Yet the 1 percent incidence rate seems to be a fairly consistent statistic worldwide. Many worthy authors have tried to shove schizophrenia into the genetic mold. It is interesting to see the lengths to which they have to go. I've read an argument comparing schizophrenics to bees.

Schizophrenics are not bees. As the shift in evolution completes its course correction over the next, oh, ten to twenty thousand years, everyone will comfortably fit into it and there will be no schizophrenia. Nor will there be uni- and bipolar illnesses or anxiety disorders I predict. As evolution distances itself from our primitive past, access to it will be less likely, and the security of dopamine suppression will be absolute. We will be farther from the seductive entropy of primitivism. However for now, while most of us hobble into the evolutionary mainstream, 1 percent don't while another 15 to 20 percent have other serious mental illnesses indicative of entropy's gravitational influence. There may be certain things that increase your odds of becoming mentally ill, but for the most part it's a function of evolution's flight path. Let's remember that a simple chemical such as LSD can bring us back to the primitive organization, at least temporarily, most likely with a brisk intensity of dopamine de-suppression triggered by the stimulation of serotonin 2A receptors. It lies within all of us and access to it is right outside our door, reconnoitered nightly in REM sleep. There is a thin line. Stressors of various kinds in certain individuals, genetic combinations, mania, and traumatic events all increase the likelihood of our opening the door or going partway there by adopting one of myriad disparate diagnoses that reflect dopamine backflow without reaching its ultimate endpoint, schizophrenia. Nonetheless, the schizophrenic 1 percent of us will roll the boulder of adult cognitivity and dopamine suppression up the hill, and just as it lodges precariously on top, the boulder will roll down to the valley below. Some things may make the hill steeper than others. Events may dislodge the boulder temporarily from its crater and vault it toward entropy's gravitational bottom. Yet the truth is, as my theory has it, overall that 1 percent is fairly fixed; all the rest just swirls around it. This represents the current state of evolution's progress in this new endeavor. It endorsed cunning after size failed the dinosaurs, and once we became verbal beings, it had to embrace the further advancement of our brains with greater complexity, dopamine suppression, distance from the primitive organization within, and loss of entropy that language and thinking demanded. Evolution moved from natural to sexual selection, survival of the fittest to survival of the most reproductive. All of these adjustments are very new in the ballet of the evolutionary time frame.

Burns (2004) offers one of the most encompassing of the so-called byproduct theories. He relates deficits in the "social brain" to schizophrenogenesis: "Schizophrenia reflects the severe end of a spectrum of abnormal cortical connectivity that appeared in our hominid ancestors approximately 150,000–100,000 years ago." He believes that the evolution of complex circuitry related to social cognition rendered the hominid brain vulnerable to genetic and environmental insults: "I believe it is appropriate to conceptualize schizophrenia as a disorder of the social brain." He talks of "progressive prolongation of brain maturation" in a theory that gets rather murky. He refers to aberrant connectivity in the social brain and states that "schizophrenia is one of the prices paid by humans for evolving complex cognitive and social abilities." This echoes Crow (2000), suggesting schizophrenia is a product of language lateralization, handedness, and a genetic speciation event. While this has merit, the reference in Burns (2004) to the social brain and connectivity tends more to obfuscate than clarify. Referring to vulnerability to insults, Burns neglects to state what the insults are to present-day schizophrenics. Ultimately he believes that schizophrenia is a genetic illness, linking the genetics to the neurodevelopment of social intuition and "mind reading" (having a feel for what others may be thinking) or meta-representation. He compares schizophrenics to autistic children.

Both Crow (2000) and Burns (2004) are wedded to genetics. Their choice of defect in schizophrenia, dysconnectivity in the social brain, or an x-linked speciation event related to laterality and handedness, die on the vine. My theory reminds us that schizophrenics are, after all, using different rules of cognitive functioning. These rules don't come from thin air. The schizophrenic's demonstrated regression of thinking to that of a child, a dreamer, an LSD user, or a primitive all demonstrate the brain's capacity in this area. A way of thought that reigned for almost six million years does not yield so easily even to the dexterous hands of evolution. With language, humans went from operating system 1.0 to operating system 2.0 under its Pied Piper–like influence, and each child repeats this progression. Their theories neglect to mention the difference between pre- and post-verbal minds and don't even begin to touch on the neurotransmitter changes that are clearly evident in schizophrenia. Language did not just result in laterality of the brain. My theory tells us where schizophrenic rules of communication come from: our past of greater than six million pre-verbal years. It explains why there is cerebral atrophy, something Burns (2004) acknowledges and attempts to fold into social brain dysconnectivity, a rather meaningless term overall. My theory defines the deterioration of the brain in terms of replacement and short-circuiting of the modern brain with a mind that functioned admirably on autopilot for millennia. My theory presents a focused redefinition of some of the core aspects of schizophrenia and explains the central paradox of its nonextinction. While Crow (2000) and Burns (2004) have merits, they are a long way from the cohesive, evidentiary model I've proposed. In its endorsement of cunning, evolution has created a shadow of the dispossessed.

One cannot have it both ways. If one affirms a genetic cause then one is stuck with fitting the round peg of schizophrenia into the Darwinian impossibility of nonextinction. Any mutation is subject to the rules of natural selection and sexual selection, and to say that linkage to other genes is what carries them forward is untenable. Too many theories crash on the genetic shoals while my theory avoids this need for intellectual contortion by asserting that schizophrenia is not a genetic process but rather an evolutionary mandate. The process that leaves 1 percent of the population worldwide with schizophrenia is the newly minted evolutionary course correction Homo sapiens find themselves in whether they recognize it or not. No other solution makes sense.

I agree, however, that social adaptations and ultimately civilizations themselves arose out of the same sweeping changes wrought by language in the last fifty thousand years or so. This state of affairs makes human interaction a key deficit of the schizophrenic trajectory since they've harkened back to a prehistoric, simpler mind set. However, it is not because of altered connectivity or mis-wired brain circuitry but precisely because the schizophrenic has regressed to a position of pre-verbality, the mental ennui of our Neanderthal neighbors, their brains having an organization extant for millions of years before language. This renders many schizophrenics cognitively devoid of a social strategy that didn't arise until Homo sapiens began to form larger groups leading to civilizations specifically as a result of our cranial forward march. What both Crow (2000) and Burns (2004) have failed to recognize is the regressive aspects of schizophrenic thought, which provides a clue to its origin. Schizophrenics don't have some derangement of language or sociableness due to genetic links to these processes that somehow carry it along.

Schizophrenic thought is an outgrowth of a specific type of regressive thinking. This is the glaring clue that links it to our past. Schizophrenics are not just genetic aberrants with inner LSD generators or sparked wiring; they are instead valiant patients with a regressed blueprint for thinking, a way of thinking that resides in all of us and that we revisit nightly in sleep. This explains the majority of the phenomenon known as psychosis.

The schizophrenic mode of communication follows a specific pattern of autistic, idiosyncratic expression with childlike rules. These rules don't provide the proper distance between the individual and the words they use, words that seem to arise syncretically from within. While they gain this talent in their this talent in their adolescence, when the burden of gated material has swollen exponentially, and as the primitive organization is also maturing, it all falls apart and entropy's hot dictum of simplicity reigns. Neither Crow (2000) nor Burns (2004) explains the age of onset of schizophrenia while my theory does, and their theories fail to explain the degenerative process seen in schizophrenia if it remains untreated. My theory, I believe, is the most encompassing, articulating answers to more questions about schizophrenia, and it is the most logical of any theory to date. It defines what schizophrenia is while definitively separating it from uni- and bipolar illnesses. It states why it is still in existence (it's not a genetic event but an evolutionary screwup); why schizophrenics communicate the way they do (regression to a primitive form of thinking); why its onset is in the late teens (the primitive organization has to mature as well and gating swells to an untenable girth for them); why antipsychotics work (the primitive organization is hyper-dopaminergic, no doubt a trait of Mr. Caveman and something children clamor to suppress); why antipsychotics lose traction (atrophy of higher centers that are short-circuited); why there is atrophy in the first place (higher centers of the brain starving for input due to replacement by the primitive organization); and where schizophrenia is going (ultimately nowhere, disappearing with the rest of mental illness as evolution gets it right millennia hence).

A true byproduct theory has to assert that schizophrenia is a result of some sideways process. My theory does just that, insisting that the demands placed on modern humans are worlds different from those before language and include neurochemical renegotiations—at least where dopamine is concerned. Not every brain is going to be pliant enough to maintain this shiny new configuration. Some will inevitably sink back to the comfort of ancient paradigms with the Pied Piper of entropy's song. As evolution completes its course correction thousands of years hence, distancing ourselves from the primitive past and far from entropy's song, 100 percent of humanity will be on board with the transformation. The mentally ill today can be viewed as evolution's dispossessed. Any change of this magnitude is likely to leave some of us behind just as the change from pre-civilized to civilized entities birthed neuroses. Ten thousand years from now or so there will be no mental illness. Humans will conform to a concatenation of normalcy and engage in processes of mental enrichment that are beyond anything we assume today. A baseline of functioning will be a given, and the mind that primitives possess will be so far behind us as to be unreachable except perhaps by the strongest psychedelic manipulations. Let us stop the *Titanic* crash on the rocks of genetics since genes, experience, and our evolutionary past all play a role in who we are.

Evolution has taken a stand against entropy and primitivity just as trees have taken a stance against gravity. Humans are in a position to leave the primitive as they mature in a way that Mr. Caveman was unable to without language. In doing so we lose entropy, increasing inner excitation as we invest energy to repress the primitive organization, suppress dopamine, and gain a higher level of cognitive organization. Entropy is the gravity that lures the boulder groundward. This price was paid in exchange for the bargain of safety, security, nutrition, and a reduced need to compete for mates. Apes have a dominance pecking order ensuring that the male leader of the group gets to breed indiscriminately with all the females. The lower-ranked male apes may be lucky enough to sneak in and go on a furtive date when the leader is not looking. Modern society embraces monogamy (although some might argue that this is breaking down), which at least guarantees the male a mate for a lifetime without having to go up against a mighty clan ruler. Over the next ten thousand years, the bar will be set higher. We will lose more entropy and gain greater mental organization, exploiting all the neurochemical legerdemain inherent in this process. We will be even farther from the disorganization of primitivity of our ancestors and presumably have less entrée to its charms. The hill we roll our boulder up as we mature will be higher and the disappointment of Sisyphus will be less likely as we distance ourselves light years from our simplistic past. Dopamine suppression will be more secure since evolution will not have just tip-toed into this new journey. We are all riding on the contextual barge of this newly minted course correction even though we don't know it. We are standing there in that giant dinosaur footprint, binoculars in hand and scanning the horizon, saying, "I see nothing." In that light, instead of attempting to fit the round peg of schizophrenia into the square hole of genetics, let's see it for what it is. The crucible of evolution has spoken; now it's our time to listen.

There are some theorists with views similar to mine. They view schizophrenia as the evolutionary drawback it surely was and continues to be. Some mention that normal brain evolution may result in abnormal connections within the brain. This random process is somewhat belied by the well-organized blueprint of brain development humans undergo. Neural misconnections would then be a regular event without cause. Others propose a tripartite structure to the brain from our past. This seems to coincide with my view that there is a primitive organization that we have access to and that can reassert itself in some individuals. Our brains contain newer and older layers that ultimately work in tandem since as the brain evolved, new structures were added to older ones. We all have access to this older scaffolding under certain circumstances, and the more primordial, diffuse, and less-structured layers have one characteristic: entropy. Entropy favors a return to the atavistic simplicity of our animalistic past and thus a rolling of the boulder back down the hill of our old friend Sisyphus. The boulder may hurtle to the ground or settle lodging in the netherworld between as is the case with depression, anxiety problems, uni- and bipolar disorders, and others. Here again we're looking at a unitary theory of mental illness that connects all forms in one symphonic performance with dopamine non-suppression the modus operandi. If that is the case, then mental illness as we know it may well vanish a mere ten or twenty thousand years hence when evolution will have moved its sweeping way to finality. The magnet of primitivity will not have the same siren intensity even if entropy yearns for it. Our minds will be so highly organized and intensely specific, with differentiation of parts and their roles, that primitivity as it is defined (a return to childlike rules of thinking and disorganization) will not be accessible. Maturation may then span some thirty years, but our lifespans may be much longer.

I agree that schizophrenia emerged in the last fifty thousand years or so when language started to flourish and that primitive forms of language were probably extant in various species of Homo before this. However, language didn't just enhance communication; it was transformative for Homo sapiens due to our superbly unique talent for suppressing dopamine, a talent that starlit our hegemony over poor Mr. Neanderthal, Denisovan, or Naledi. It is as if the species were hungry for this metamorphosis, just waiting for the clown car of its transformation. Perhaps, with such a possibility, the inklings of speech were seized on by Homo sapiens and, like a lock and key, they opened the door to modern humans that the other species of Homo just could not enter. An evolutionary gradient was teed up in which those pre-sapiens who could suppress dopamine even slightly lurched forward until some beaming luck hit a mutational homer that propelled a sapiens into this granite spotlight. This transformative genie enriched the mind of Homo sapiens, fertilizing conscious thought, empowering us to solve problems that were unapproachable in past eras, gating out irrelevant stimuli and hardening our egos. Evolution had made its choice after the wreck of dinosaur bones that sank under their own weight in an impetuous paradigm of size superiority. It chose cunning, and cunning led to brain expansion with a flowering of the prefrontal cortex, and when the critical mass and configuration was reached, the explosion shook the universe. We were able to fashion the world Mother Nature bestowed upon us in our own uniqueness, pave the earth, string electricity across the pine-treed land, and flick on an electric light bulb. And when Homo sapiens sat back and stared out over what they had created, they could see the lengths they'd gone to abandon lanky primitivity, the primitive disorganization that once was what they were. Staring across the horizon, we can see the positives and the negatives.

The next wave will have smarter, more efficient brains and will again, no doubt, reorganize the earth in their fashion. This is only the first step. We are still in the process of evolution's searing torque.

* * *

Some time ago I was lounging in a small diner, swirling my favorite hot chocolate powder into a cup of steamed soy milk and gazing blandly out the window. The paved roads, honking cars, and stoplights all reminded me that one million years past this was all jungle. A Homo sapiens may have hobbled across this spot dressed in animal skins, desperate for food and willing, if necessary to eat his own kind, or any animal that crossed his path if he could. His condo was a cave and the mirror was a placid pond. Knowing no nomenclature, he groveled now on two legs not four through his inchoate existence, desperate to survive and often folding in that brutal gamble. I took a sip of hot chocolate and realized it required agriculture to produce the cacao beans, machinery to grind them, kettles to roast them, and porcelain to make the cup. Half a million years ago we did not have those things nor did we have them ten thousand years past. What happened to us, the noble human of nature, scratching his way through an existence that often met an ignominious end?

Somebody somewhere made some kind of bargain, did they not? And Mother Nature got the short end of the stick. Mankind, on the other hand, a part of nature, no longer faced the excruciation of the jungle and we careened from cave to condo in the past three thousand years or so, setting foot on the moon. There is evidence of cannibalism in our ancestors, and it is understandable, I guess, in the thick of starvation to eat one's own, the goal of survival holding priority. We take for granted certain things like air conditioning, universities, and the Eiffel Tower. Having let my drink cool a bit, I took a sip. It never fails to entice me. I bite down on a piece of chocolate and chew letting it captivate my taste buds. Who am I? Nowadays we can hire a company to provide a tour guide to lead us on a trip to what once was our habitat, and we can fly on a jet to a place that will take us on a boat to the Galapagos Islands where Darwin made his amazing discoveries and we can read an e-book on the dangers of pesticides. None of that existed, oh, three hundred years ago.

"We paved paradise and put up a parking lot," Joni Mitchell once sang. But who can blame us? If given a choice between living unprotected or residing in a suburban, cookie-cutter community, who would choose the former? Malvina Reynolds wrote a song about it: "Little Boxes." Maybe the houses are all the same, but they keep us warm in winter and cool in summer's scorch. We chose the obvious, and it all came down to safety and survival. It was barely even a conscious choice. We did it out of necessity and convenience combined and took a plain and made it a community. Was any of this planned? No, not really. It just happened.

But it couldn't have happened before language. Language was the ayahuasca we needed at the moment, and now look. It transformed our brains, and then we transformed our world. Homo sapiens were ready, brain dripping in anticipation of that elephantine gift. We seized upon the crumbs of whatever primitive form of speech was happening and vaulted skyward with it. Little did we know back then what the result would be, that it would change us forever so that we could stand out of sweet Mother Nature's geodome and pave the earth. Not only would we live differently, but we would think differently and our minds would function on a crisper level. Who knew? All because some caveperson saw berries on a bush and said, "Skus." We've stumbled into all this gold like half-blind, tipsy prospectors tripping over the mother lode. We built monuments in the form of skyscrapers without even considering who we are and built zoos so we can gaze perplexedly through iron bars at what we used to be. There's no conductor although the religious would assert that there certainly is.

But a metamorphosis of this magnitude may not include everybody at first blush. One size does not fit all. Schizophrenics and all the mentally ill are evolution's dispossessed. If the entire mass of humanity's streaming protoplasm had to make a change, it could not be expected to be a seamless one, and it is one that teaches us who we are. We boarded the mother ship of language but we're still close enough to primitivity and Planet Entropy's gravity so that 1 percent get sucked back to that orb. We are all very brief flashes in time, flaming out quickly as we make our tiny contribution to the forward thrust of protoplasm, hopefully donating our genes to the future. Our children are the new and improved us, and their children better yet. Humankind ten thousand years from now would be barely recognizable to us.

Suddenly I felt something move, shifting around me. There was a swaying motion and I saw the shore begin to drift away. I remembered that I was not in a diner after all but rather on an elite ocean liner. I'd booked a cruise to the Bahamas and the ship was pulling away from the dock. People on shore were waving to their loved ones, and I was in something much bigger than myself. We were leaving them behind, and as we did so, I wondered why everyone couldn't just join us. It was to be an eight-day cruise with a lengthy stopover. Some, who had been on board visiting, were now back onshore, left out. They could only wave and regret that they couldn't mingle with the rest of us. I looked at them and wondered if I could find some way to lift them from shore and smuggle them onto the boat. They seemed like perfectly reasonable people. "C'mon," I said, "hop on some dinghy and motor over."

They couldn't hear me.

Notes

Chapter 7

1. There are people who used to be called "idiot savants." They can do one thing outstandingly well. My theory is that they're doing it without the benefit of secondary process thinking 2.0, which they sidestep in the landscape of a language deficit. They have found a way to achieve incredible results, let's say in mathematics, without venturing into the conceptual realm mentally. They don't see other humans as sources of gratification, and they fastidiously sidestep that most indelible of dopamine suppression's talents, the gating function, which around age five plunges the primitive organization into an unconscious vault of darkness, a feat requiring significant energy.

2. Hallucinogen users suffer an ego dissolution that's temporary in most. Thus when DMT in ayahuasca triggers the serotonin 2A receptor, dopamine is de-suppressed like a freight train, dissolving the Captain Kirk ego and its sharp boundaries, returning to primitive thinking with hallucinations.

3. The success of lobotomy was probably due to the dismantling and disconnection of these higher brain structures, allowing the more primitive ones to take unfettered control, thereby eliminating any conflict between the two. In one series, 83 percent were improved with orbitofrontal lobotomy (Green et al. 1952).

Chapter 8

1. We then test these hypotheses with rational experimentation. The standard experiment will be double blind, randomized, and placebo controlled with plenty of subjects to test the issue we are trying to resolve because more subjects give the experiment more power. What does that mean? We assume that certain rules make an experiment more viable and rigorous. "Double blind" means that both the raters of the response and the subjects, the ones who get the experimental drug, do not know who has been given the drug and who hasn't. If the raters knew who had been given the drug, they would be more likely to rate a subject as improved even if they were trying not to do this. If the subject knows they have been give the experimental drug, they are more likely to feel better in the way a placebo causes people to improve. It's interesting to note that placebo response in psychiatry can be upward of 30 percent. If you give one hundred people a sugar pill and

say, "This is going to make your depression better," about thirty of them will feel better. Placebos can be very powerful.

The U.S. Food and Drug Administration generally requires two double-blind, placebo-controlled studies proving a product is effective. The analytic standard is 0.05, meaning that ninety-five times out of one hundred you can expect to get the same result.

Finally, "randomized" means that the subjects getting the medication and those getting the placebo are chosen at random. If they were handpicked by the experimenters, the experimenters could choose people with a certain characteristic that made them more likely to respond to the drug. Papers often give a comparison list of demographics of the placebo group versus the experimental group.

2. Suicidal methods vary considerably in their effectiveness. Some are very likely to fail; others almost guaranteed to work. Without naming which is which, it is true that women make more attempts, that men succeed more often, and that this is based on the choice of which means men and women use. Thank God most people choose less-lethal means. Perhaps this indicates an unconscious wish to live.

3. Does this remind you of anything? Remember our discussion of brain laterality? The brain enlarged on the left side to accommodate language and resulted in handedness, usually right-handedness. Before this laterality there was no handedness. One way anthropologists check a fossilized skull for handedness is to look at the teeth. Apparently Mr. Caveman used his teeth to cut things. Holding whatever it was in his mouth, he pulled it with one hand and whacked it with a stone tool with the other. Based on markings on fossilized teeth, anthropologists can tell if the individual was right- or left-handed or ambidextrous. Before language there probably was no handedness. Now remember our discussion of the D2:D4 finger ratio? Schizophrenics with Schneiderian symptoms have smaller D2:D4 ratios, which seems to be related to laterality of the brain, which is related to language and handedness.

4. Once again I point out the dichotomy between primary process thinking and secondary process. Manics are stuck in primary process thinking but with the accelerator pushed to the floor. Remember our tree branch analogy? Manics, barely hanging on to the branch, are doused in a raging river of dopamine. Sometimes they let go and become quite psychotic.

References

Please read for yourself these fabulous books by people of various disciplines who have made it their business to enlighten and educate us. You managed to wade through this book, and you will find some of these harder, others easier. They are all very worthy. There are so many more sources that are deserving of your time—the intrapsychic adventurer needs to traverse fields as diverse as psychoanalysis, thermodynamics, neurochemistry, anthropology, Darwinian evolution, cartoons (remember the Charles Addams footprint?), and of course schizophrenia and mental illness. There are many excellent books written by sufferers of mental illness who show not only talent but bravery in their self-exposure and wisdom. Do yourself a big favor and read some. Oh, what language has wrought! But my enthusiasm for all of this orchestrated extravaganza remains even when the espresso steamer is nowhere in sight. I hope I've imparted a little of it to you.

Berlim, M. T. "The Etiology of Schizophrenia and the Origin of Language: Overview of a Theory." *Comprehensive Psychiatry* 44, no. 1 (2003): 7–14.

Boland, E. M., et al. "Meta-analysis of the Antidepressant Effects of Acute Sleep Deprivation." *Journal of Clinical Psychiatry* 78, no. 8 (2017).

Bolu, A., et al. "The Ratios of 2nd to 4th Digit May Be a Predictor of Schizophrenia in Male Patients." *Clinical Anatomy* 28 (2015): 551–56.

Brebion, G., et al. "Verbal Fluency in Male and Female Schizophrenia Patients: Different Patterns of Association with Processing Speed, Working Memory Span, and Clinical Symptoms." *Neuropsychology* 32, no. 1 (2018): 65–76.

Brune, M. "Schizophrenia—An Evolutionary Enigma?" *Neuroscience and Biobehavioral Reviews* 28, no. 1 (2004): 41–53.

Burns, J. K. "An Evolutionary Theory of Schizophrenia: Cortical Connectivity, Metarepresentation, and the Social Brain." *Behavioral and Brain Sciences* 27, no. 6 (2004): 831–85.

Cahn, W., et al. "Brain Volume Changes in First-Episode Schizophrenia: A 1-Year Follow-up Study." *Archives of General Psychiatry* 59, no. 22 (2002): 1002–10.

Callier, S., et al. "Evolution and Cell Biology of Dopamine Receptors in Vertebrates." *Biology of the Cell* 95, no. 7 (2003): 489–502.

Cardno, A., and I. Gottesman. "Twin Studies of Schizophrenia: From Bow-and-Arrow Concordances to Star Wars Mx and Functional Genomics." *American Journal of Medical Genetics* 97, no. 1 (2000): 12–17.

Chabot, C. B. *Freud on Schreber: Psychoanalytic Theory and the Critical Act*. Amherst: University of Massachusetts Press, 1982.

Commons, K. G., and S. E. Linnros. "Delayed Antidepressant Efficacy and the Desensitization Hypothesis." *ACS Chemical Neuroscience* 10, no. 7 (2019): 3048–52.

Corballis, M. C. *The Truth about Language: What It Is and Where It Came From*. Chicago: University of Chicago Press, 2017.

Crow, T. J. "March 27, 1827 and What Happened Later—The Impact of Psychiatry on Evolutionary Theory." *Progress in Neuro-Psychopharmacology and Biological Psychiatry* 30, no. 5 (2006): 785–96.

———. "The Missing Genes: What Happened to the Heritability of Psychiatric Disorders?" *Molecular Psychiatry* 16 (2011): 362–64.

———. "Schizophrenia as the Price that Homo Sapiens Pays for Language: A Resolution of the Central Paradox in the Origin of the Species." *Brain Research Reviews* 31, nos. 2–3 (2000): 118–29.

———. "Schizophrenia as Variation in the Sapiens-Specific Epigenetic Instruction to the Embryo." *Clinical Genetics* 81, no. 4 (2012): 319–24.

Dani, J. A. "Neuronal Nicotinic Acetylcholine Receptor Structure and Function and Response to Nicotine." *International Review of Neurobiology* 124 (2015): 3–19.

De Gregorio, D. et al. "d-Lysergic Acid Diethylamide (LSD) as a Model of Psychosis: Mechanism of Action and Pharmacology." *International Journal of Molecular Sciences* 17, no. 11 (2016): 1953–73.

Everett, D. L. *How Language Began: The Story of Humanity's Greatest Invention*. New York: W. W. Norton & Company, 2017.

Ferre, S. "Mechanisms of the Psycho-Stimulant Effects of Caffeine: Implications for Substance Use Disorders." *Psychopharmacology* 233 (2016): 1963–79.

Ferre, S. et al. "New Developments on the Adenosine Mechanisms of the Central Effects of Caffeine and Their Implications for Neuropsychiatric Disorders." *Journal of Caffeine and Adenosine Research* 8, no. 4 (2018): 121–31.

Freud, S. *Beyond the Pleasure Principle*. Edited by James Strachey. New York: W. W. Norton & Company, 1961.

———. *The Ego and the Id*. New York: W. W. Norton & Company, 1960.

———. *A General Selection from the Works of Sigmund Freud*. Edited by John Rickman. Garden City, NY: Doubleday, 1957.

———. *Three Case Histories*. Edited by Philip Rieff. New York: Macmillan, 1963.

Gebhardt, J., et al. "Maturation of Prepulse Inhibition (PPI) in Childhood." *Psychophysiology* 49, no. 4 (2012): 484–88.

Goff, D. C. "Longer Duration of Untreated Psychosis Linked to Loss of Brain Volume." *JAMA Psychiatry* (February 2018).

Green, J. R., et al. "Orbitofrontal Lobotomy with Reference to Effects on 55 Psychotic Patients." Presented at the joint meetings of the San Francisco Neurological, Southern California Neurosurgical and Western Electroencephalographic Societies, Del Monte Lodge, Pebble Beach, California, 1952.

Hacksell, U., et al. "On the Discovery and Development of Pimavanserin: A Novel Drug Candidate for Parkinson's Psychosis." *Neurochemical Research* 39, no. 10 (2014): 2008–17.

Harari, Y. N. *Sapiens: A Brief History of Humankind*. New York: HarperCollins, 2015.

Heal, D. J. et al. "Amphetamine, Past and Present: A Pharmacological and Clinical Perspective." *Journal of Psychopharmacology* 27, no. 6 (2013): 479–96.

Hofman, M. A. "Evolution of the Human Brain: When Bigger Is Better." *Frontiers in Neuroanatomy* 8 (2014): 1–12.

Howells, J. G., ed. *The Concept of Schizophrenia: Historical Perspectives*. Washington, DC: American Psychiatric Association, 1991.

Huber, R. et al. "Human Cortical Excitability Increases with Time Awake." *Cerebral Cortex* 23, no. 2 (2013): 332–38.

Joseph, J. "Don Jackson's 'A Critique of the Literature on the Genetics of Schizophrenia': A Reappraisal after 40 Years." *Genetic Social and General Psychology Monographs* 127, no. 1 (2001): 27–57.

Kalmady, S. V., et al. "Relationship between Brain-Derived Neurotrophic Factor and Schneiderian First Rank Symptoms in Antipsychotic-Naive Schizophrenia." *Frontiers in Psychiatry* 4, no. 64 (2013).

Klein, M. O., et al. "Dopamine: Functions, Signaling, and Associations with Neurological Diseases." *Cellular and Molecular Neurobiology* 39, no. 1 (2019): 31–59.

Leakey, R. *The Origin of Humankind*. New York: Basic Books, 1994.

LeDoux, J. *The Deep History of Ourselves: The Four-Billion-Year Story of How We Got Conscious Brains*. New York: Penguin Books, 2020.

Lee, S. H., and S. Y. Yoon. *Close Encounters with Humankind: A Paleoanthropologist Investigates Our Evolving Species*. New York: W. W. Norton & Company, 2018.

Lerario, A. et al. "Charles Bonnet Syndrome: Two Case Reports and Review of the Literature." *Journal of Neurology* 260, no. 4 (2013): 1180–86.

Lesk, S. "A Different View of Patients with Schizophrenia." *Current Psychiatry* 17 (2018): e1–e3.

———. "Taking Acid." *New York Times Book Review*, June 24, 2018.

Llado-Pelfort, L., et al. "Effects of Hallucinogens on Neuronal Activity." *Current Topics in Behavioral Neurosciences* 36 (2018): 75–105.

Lothane, Z. "The Teachings of Honorary Professor of Psychiatry Daniel Paul Schreber, J.D., to Psychiatrists and Psychoanalysts, or Dramatology's Challenge to Psychiatry and Psychoanalysis." *Psychoanalytic Review* 98, no. 6 (2011): 775–815.

Lowery, R. K., et al. "Neanderthal and Denisova Genetic Affinities with Contemporary Humans: Introgression versus Common Ancestral Polymorphisms." *Gene* 530, no. 1 (2013): 83–94.

Mansbach, R. S., et al. "Blockade of Drug-Induced Deficits in Prepulse Inhibition of Acoustic Startle by Ziprasidone." *Pharmacology, Biochemistry, and Behavior* 69, nos. 3–4 (2001): 535–42.

Martiny, K. "Novel Augmentation Strategies in Major Depression." *Danish Medical Journal* 64, no. 4 (2017).

Moser, D. A., et al. "Multivariate Analyses Reveal Link between Brain Variance, Psychosis." *JAMA Psychiatry* (2017).

Niederland, W. G. *The Schreber Case: Psychoanalytic Profile of a Paranoid Personality*. New York: Quadrangle Books, 1974.

Ocklenburg, S. et al. "Laterality and Mental Disorders in the Postgenomic Age—A Closer Look at Schizophrenia and Language Lateralization." *Neuroscience and Biobehavioral Reviews* 59 (2015): 100–110.

Ostler, T. "Heinz Werner: His Life, Ideas, and Contributions to Developmental Psychology in the First Half of the 20th Century." *Journal of Genetic Psychology* 177, no. 6 (2016): 231–43.

Pagel, M. D. *Wired for Culture: Origins of the Human Social Mind*. New York: W. W. Norton & Company, 2012.

Pearlson, G. D., and B. S. Folley. "Schizophrenia, Psychiatric Genetics, and Darwinian Psychiatry: An Evolutionary Framework." *Schizophrenia Bulletin* 34, no. 4 (2008): 722–33.

Polimeni, J., et al. "Evolutionary Perspectives on Schizophrenia." *Canadian Journal of Psychiatry* 48, no. 1 (2003): 34–39.

Power, R. A., et al. "Fecundity of Patients with Schizophrenia, Autism, Bipolar Disorder, Depression, Anorexia Nervosa, or Substance Abuse vs. Their Unaffected Siblings." *JAMA Psychiatry* 70, no. 1 (2013): 22–30.

Qi, H., and S. Li. "Dose-Response Meta-analysis on Coffee, Tea and Caffeine Consumption with Risk of Parkinson's Disease." *Geriatrics and Gerontology International* 14, no. 2 (2014): 430–39.

Riga, M. S., et al. "The Serotonin Hallucinogen 5-MeO-DMT Alters Cortico-Thalamic Activity in Freely Moving Mice: Regionally-Selective Involvement of 5-HT_{1A} and 5-HT_{2A} Receptors." *Neuropharmacology* 142 (2018): 219–30.

Rudan, I. "New Technologies Provide Insights into Genetic Basis of Psychiatric Disorders and Explain Their Co-morbidity." *Psychiatria Danubia* 22, no. 2 (2010): 190–92.

Sacheli, M. A., et al. "Habitual Exercisers versus Sedentary Subjects with Parkinson's Disease: Multimodal PET and fMRI Study: Exercising versus Sedentary Subjects with PD." *Movement Disorders* 33, no. 3 (2018).

Saks, E. R. *The Center Cannot Hold: My Journey through Madness*. New York: Hachette Book Group, 2007.

Satoh, H., et al. "Downregulation of Dopamine D1-like Receptor Pathways of GABAergic Interneurons in the Anterior Cingulate Cortex of Spontaneously Hypertensive Rats." *Neuroscience* 394 (2018): 267–85.

Schatzman, M. "Paranoia of Persecution: The Case of Schreber." *International Journal of Psychiatry* 10 (1972): 53–78.

Schreber, D. P. *Memoirs of My Nervous Illness*. Cambridge, MA: Harvard University Press, 1988.
Seeman, P. "Cannabidiol Is a Partial Agonist at Dopamine D2High Receptors, Predicting Its Antipsychotic Clinical Dose." *Translational Psychiatry* 6, no. 10 (2016): 1–4.
Srinivasan, S., et al. "Genetic Markers of Human Evolution Are Enriched in Schizophrenia." *Biological Psychiatry* 80, no. 4 (2016): 284–92.
Stabenau, J. R., and W. Pollin. "Heredity and Environment in Schizophrenia, Revisited: The Contribution of Twin and High-Risk Studies." *Journal of Nervous and Mental Disease* 181 (1993): 290–97.
Sullivan, P. F., et al. "Schizophrenia as a Complex Trait: Evidence from a Meta-analysis of Twin Studies." *Archives of General Psychiatry* 60, no. 12 (2003): 1187–92.
Swerdlow, N. R., et al. "Startle Gating Deficits in a Large Cohort of Patients with Schizophrenia: Relationship to Medications, Symptoms, Neurocognition, and Level of Function." *Archives of General Psychiatry* 63 (2006): 1325–35.
Takahashi, H., et al. "Prepulse Inhibition of Startle Response: Recent Advances in Human Studies of Psychiatric Disease." *Clinical Psychopharmacology and Neuroscience* 9, no. 3 (2011): 102–11.
Tausk, V., and D. Feigenbaum. "On the Origin of the 'Influencing Machine' in Schizophrenia." *Journal of Psychotherapy Practice and Research* 1, no. 2 (1992): 184–206.
Tikka, S. K., et al. "Schneiderian First Rank Symptoms in Schizophrenia: A Developmental Neuroscience Evaluation." *International Journal of Developmental Neuroscience* 50 (2016): 39–46.
Trautmann, N., et al. "Response to Therapeutic Sleep Deprivation: A Naturalistic Study of Clinical and Genetic Factors and Post-treatment Depressive Symptom Trajectory." *Neuropsychopharmacology* 43, no. 13 (2018): 2572–77.
Uher, R. "Gene-Environment Interactions in Severe Mental Illness." *Frontiers in Psychiatry* 5 (2014): 48.
———. "The Role of Genetic Variation in the Causation of Mental Illness: An Evolution-Informed Framework." *Molecular Psychiatry* 14 (2009): 1072–82.
Venkatasubramanian, G., et al. "Digit Ratio (2D:4D) Asymmetry and Schneiderian First Rank Symptoms: Implications for Cerebral Lateralisation Theories of Schizophrenia." *Laterality* 16, no. 4 (2011): 499–512.
Vernier, P., et al. "Bioamine Receptors: Evolutionary and Functional Variations of a Structural Leitmotiv." In *Comparative Molecular Biology*, edited by Y. Pichon, 297–337. Basel: Birkhauser Verlag, 1993.
———. "The Degeneration of Dopamine Neurons in Parkinson's Disease: Insights from the Embryology and Evolution of the Mesostriatocortical System." *Annals of the New York Academy of Sciences* 1035 (2004): 231–49.
Wang, E. W. *The Collected Schizophrenias*. Minneapolis, MN: Graywolf Press, 2019.
Wearne, T. A., and J. L. Cornish. "A Comparison of Methamphetamine-Induced Psychosis and Schizophrenia: A Review of Positive, Negative, and Cognitive Symptomatology." *Frontiers in Psychiatry* 9 (2018): 491.
Werner, H. *Comparative Psychology of Mental Development*. New York: International Universities Press, 1948.

Werner, H., and B. Kaplan. *Symbol Formation: An Organismic-Developmental Approach to Language and the Expression of Thought*. New York: John Wiley & Sons, 1963.

Wolf, E., et al. "Synaptic Plasticity Model of Therapeutic Sleep Deprivation in Major Depression." *Sleep Medicine Reviews* 30 (2016): 53–62.

Yeiser, B. *Mind Estranged: My Journey from Schizophrenia and Homelessness to Recovery*. Create Space, 2014.

Zant, J. C., et al. "Increases in Extracellular Serotonin and Dopamine Metabolite Levels in the Basal Forebrain During Sleep Deprivation." *Brain Research* 1399 (2011): 40–48.

Zimmer, C. *Smithsonian Intimate Guide to Human Origins*. New York: Harper Perennial, 2007.

Index

Dicsclaimer

All content and information in this book is for informational and educational purposes only. Much of this book is theoretical. Nothing in this book constitutes medical advice. Nothing in this book is intended to be a substitute for professional medical advice, diagnosis, or treatment. Always seek the advice of your physician or other qualified health care provider with any questions you may have regarding a medical condition or treatment and before undertaking a new health care regimen. Never disregard professional medical advice or delay in seeking it because of something you have read in this book.